CLOUDS

CLOUDS

Eric M. Wilcox

Foreword by Gavin Pretor-Pinney

DUNCAN BAIRD PUBLISHERS

LONDON

For Caroline—our Bright Star

Clouds
Eric M. Wilcox

First published in the United Kingdom
    and Ireland in 2008 by
Duncan Baird Publishers Ltd
Sixth Floor, Castle House
75–76 Wells Street
London W1T 3QH

Produced for Duncan Baird Publishers by
Herter Studio LLC
432 Elizabeth Street
San Francisco CA 94114
USA

Managing Editor: Christopher Westhorp
Editor: Susie Hogarth
Managing Designer: Daniel Sturges
Designer: Debbie Berne
Picture Manager: Julia Brown
Commissioned artwork: Sailesh Patel

British Library Cataloguing-in-Publication
Data:
A CIP record for this book is available
from the British Library

ISBN: 978-1-84483-688-8

10 9 8 7 6 5 4 3 2 1

Typeset in Filosofia and Gill Sans
Colour reproduction by Scanhouse,
    Malaysia
Printed in Singapore by Imago

Captions to illustrations on pages 2–9
pages 2–3
A hazy pileus cloud caps
a rising cumulus cloud.

pages 6–7
Altocumulus lenticularis
beneath cirrostratus.

pages 8–9
Fall streaks of ice crystals
known as cirrus uncinus.

TABLE OF CONTENTS

FOREWORD

The first time I noticed a cloud was when I was four and a half, looking from the window of my mother's car as she drove me to school. The cloud was obscuring the sun so that its crisp cauliflower mounds were fringed with dazzling white and dramatic shafts of sunlight burst out from behind it. This fan of rays must have drawn my attention to the cloud in the first place. I wondered at that moment what a cloud was made of, why it was up there, and what would it feel like to sit on?

Every child seems to go through a cloud phase, during which they become fascinated and intrigued by them. After all, as bodies without surfaces, they are rather perplexing things. Their shifting, amorphous forms are also the perfect inspiration for active imaginations. Who can't recall lying back and gazing at the sky to pick out shapes in the clouds: fantastical animals, grotesque faces, strange upside-down landscapes? "The appearances of nature are the truths of nature," wrote the Victorian author, George Macdonald. "It is through their show, not their analysis, that we enter into their deepest truths. What they say to the childlike soul is the truest thing to be gathered of them."

The glorious photographs in this book serve as reminders that clouds are the most varied and magnificent of nature's displays. Perhaps more than any others, they can also have a profound effect on how we feel. Low, overcast skies hang oppressively, making us feel claustrophobic, "under the weather". Woolly, fair-weather clouds have a light-hearted flavour, as they lazily flock across a sunny afternoon. From a distance, immense storm clouds bring a humbling sense of scale to our ocean of air. From down below, their wild, elemental power is both invigorating and life affirming. The delicate cascading ice crystals of the high clouds form weightless, wispy streaks that can't fail to lift the spirits. Clouds pass over us like moods.

John Constable, perhaps Britain's most accomplished cloud painter, believed "we see nothing truly till we understand it". I think that it pays to learn to recognize and understand some of the many and varied formations that clouds adopt. Each is like an expression on the face of the atmosphere, revealing the capricious confluence and divergence of its invisible currents; sometimes presaging a turn in the weather to come. While the cloud photographs in this book can be appreciated for their own sake, their beauty is most definitely enhanced by Eric Wilcox's lucid explanations and informative commentary.

It is a shame that as people grow accustomed to the omnipresence of clouds they become blind to them or see them merely in terms of their physical outcome: no more than so many inches of rain, sleet or snow. Have you ever noticed the way stressed people look to the ground? Contemplating the clouds opens your chest like a yoga position. The clouds move so subtly that you have to slow down to appreciate them. Ralph Waldo Emerson described the sky as "the daily bread of the eyes".

I'm sure that the images and explanations in this book will help to open eyes to our ever-changing, ever-surprising cloudscapes. Cloud-gazing is a gloriously aimless pursuit. And this is why, as much as anything, it is such a profoundly valuable thing to do.

Gavin Pretor-Pinney

GAVIN PRETOR-PINNEY

Founder of the Cloud Appreciation Society

www.cloudappreciationsociety.org

INTRODUCTION

It should be no mystery why humans have always harbored an obsession with the weather. Our ability to grow food, our access to drinking water, and our safe navigation of the seas are just a few of the critical human endeavors whose success or failure depends on the weather—or perhaps more precisely on our ability to predict and mitigate the development of adverse weather. But what explains our fascination with clouds?

Deceptively simple in composition, each cloud is merely a collection of tiny drops of water and crystals of ice. Yet clouds are obviously so much more than that. Out of such elementary constituents come a multitude of shapes, and skyscapes that range from the serene to the tumultuous. With an ability to create elaborate patterns from the simplest of components, clouds are a perfect example of nature's complexity. Collisions of tiny drops of water conspire with continental-scale patterns of wind to produce expansive curtains of fibrous cirrus, endless fields of puffy cumulus, and violent towers of cumulonimbus. The drama of these exhibitions plays out daily above our heads, and we are all free to partake of its grandeur.

Certainly there is a practical element to sky-watching. Centuries of weather lore have provided us with warnings such as "an evening grey and a morning red, will send the shepherd wet to bed"—a reference to approaching clouds from the west, which are painted red at sunrise, as an indicator of an approaching storm. But clearly there are deeper connections between clouds and our daily experiences and moods. Compare the jubilation of a bright blue sky dotted with fluffy buoyant clouds to the pall of a drizzle beneath a low grey ceiling. Aristophanes referred to clouds as "the patron goddesses of idle men", presumably a reference to the enduring, and languorous, pastime of cloud-gazing.

CLASSIFYING AND CELEBRATING CLOUDS

While the pleasures of contemplating the varied shapes of clouds have been appreciated by countless generations, a suitable scheme for classifying cloud formations was not achieved until the publication in 1803 of Luke Howard's *Essay on the Modification of Clouds*. In it, Howard provides the Latin terminology, including "cumulus" (from the Latin for "heap", to describe the tall billowing clouds), "cirrus" (from the Latin for "hair", to describe the thin fibrous clouds), "stratus" (from the Latin for "layer", to describe the layered clouds), and "nimbus" (from the Latin for "cloud", to describe the raining clouds), that persists today in our modern classification of clouds.

The mercurial nature of clouds has inspired their use in centuries of literature as a metaphor for changing moods and crises of identity. Such transience in form and appearance poses a particular challenge for classifying clouds. Howard's great accomplishment was the recognition that clouds could be classified according to their appearance, even while their form frequently changes—even from one cloud type to another. For this reason, Howard referred to the cloud types as "modifications", describing how cloud form relates to the dynamics of meteorology, as well as describing common progressions of related cloud types. These insights helped pave the way for the tremendous progress in understanding meteorology that followed in the nineteenth and twentieth centuries.

Lest one be concerned that such attempts to describe clouds scientifically act to dilute the wonder and majesty of the clouds themselves, it is important to note that Luke Howard's treatise rode in on the coattails of the enlightenment, and true to the period, inspired a number of artists and intellectuals. The poets Goethe and Shelley were both contemporaries of Howard and they created works that drew directly from imagery of his cloud types. The English painter John Constable, a self-described "man of clouds", made a close study of Howard's work while developing his own study of clouds in a famous collection of skyscapes and cloudscapes.

FORM AND FORECASTING

Meteorology is both a young and a very old field of study. It is no surprise that weather forecasts are found in many ancient texts. Just as the weather makes a frequent topic of conversation in the modern world, so it must have centuries and millennia before. And just as the ancients lived immersed in and bound by their environment, we in the modern world are no less dependent for our existence on the fresh water supplied to us by clouds, no matter how much many of us are insulated from nature's harshest extremes or even its daily nuisances. Modern technology has allowed us to generate weather forecasts days in advance, yet clouds somehow remain fraught with mystery. A small army of scientists still probes the curious behavior of clouds, slowly working out the details of cloud formations but still struggling to understand how to predict exactly when and where clouds will form.

Although the roles of some basic mechanisms of cloud formation, such as buoyancy, were understood at the time of Howard's treatise on clouds, the details of the entire process of cloud formation have proven themselves to be highly complex and vexing. There remains in existence to this day an active, worldwide community of cloud physicists, as well as a burgeoning community of cloud chemists. The extreme range of scales at work in clouds reflects the enormity of the challenge—from the growth of the smallest drops that are only tiny fractions of an inch in diameter, to the wind patterns that organize cloud systems hundreds of miles in diameter, and all manner of processes in between.

Among the important consequences of clouds still under investigation today is the effect of clouds on the climate of Earth. Clouds, it is now understood, have both a cooling effect, by casting shadows over the air and ground beneath, and a warming effect by reducing the rate at which the surface and lower atmosphere radiate energy to space to balance the heat absorbed from the Sun (the so-called "cloud greenhouse effect"). Modern measurements from satellites have revealed that the net result of these two cloud effects is to cool the planet; that is, Earth is substantially cooler today than it would be if the clouds were suddenly taken from us. However, the extent to which cloud cover has changed in the past, with the waxing and waning of Earth's climate from ice age to interglacial periods, remains a mystery. Likewise, the future of cloud cover remains unknown in a world whose climate is now understood to be significantly influenced by modern industrial man.

In spite of our modern conveniences, and their ability to shield us from the great variability of weather, it is heartening to know that clouds still provide a visual guide to the changing moods of the sky; that their varied forms still spark our imaginations; and that their inner machinations still provide mysteries to be unravelled. Their presence is both an opportunity to marvel at nature's majesty and an invitation to let our imaginations wander.

CLOUDS

CUMULUS

Who can really say they have looked up to see a motley assortment of cumulus clouds drifting by and not imagined what their shapes represent? The piles of fluffy cloud and well-defined boundaries of cumulus seem capable of assuming an infinite variety of shapes. Their tops appear to bulge, while appendages protrude from their sides. The cunning deception of cumulus clouds is their tendency to appear stable even while adopting lively and organic forms. The evolution of a cumulus cloud is just slow enough that, while it may appear unchanging when observed, it will have adopted a markedly different form should you divert your attention for a few moments.

"Cumulus" means "heaped" in Latin. The heaps of billowing cloud result from the tendency for the air in cumulus clouds to rise vertically in plumes. "Warm air rises," is the rule of thumb we learn as children, and indeed a cumulus cloud will continue to grow so long as the air within the cloud remains warmer than its surroundings, retaining sufficient moisture to continue creating water drops. But for the humble cumulus humilis—the shallow clouds whose shapes fuel our imagination on a fair-weather day—buoyancy is fleeting; the clouds do not have to rise far before encountering a warm layer of stable air, a layer that puts a lid on the bubbling turbulence of the lower atmosphere.

Cumulus humilis clouds are the lowest of the cumulus cloud species. In a thicker layer of instability other cumulus clouds will grow taller. Cumulus mediocris are of moderate height and will appear as tall as, or taller than, they are wide. Cumulus congestus are the tallest of the cumulus. They are usually shallow enough that their tops remain warmer than the freezing point of water, and therefore lack snow or ice; any taller and they would become cumulonimbus.. However, they may occasionally be deep enough to cause a rain shower underneath.

Upon spying a modest cumulus cloud, one might pause briefly to wonder which creature that cloud will come to resemble next. But because the buoyant action that molds the shape of the cloud also results in its growth, it might be worth watching a while longer to see if that humble cumulus humilis soars into a majestic congestus, or perhaps something even taller.

"Fair-weather" cumulus clouds embellish a blue sky. As winds blow over the Earth's surface, turbulence stirs the air above and one of the results of this mixing of the lower atmosphere is small cumulus clouds.

The steady trade winds in the tropics are frequently dotted with sheets of fair-weather cumulus like these. Above the layer of clouds the air is slowly subsiding, ensuring that these clouds remain small. Low clouds are common over the ocean, even in fair-weather conditions, because of the abundant moisture available from the sea.

Cumulus humilis clouds are generally wider than they are tall. These examples display the classic form of a flat bottom and rounded tops. However, the fact that countless dreamers have stared at cumulus clouds wondering what their shapes resemble is a testament to the remarkable number of variations that there are to this form.

Cumulus clouds may remain shallow if there is insufficient moisture to continue making cloud drops, or if the rising cloud-top encounters a stable warm layer. It is impossible to tell whether these cumulus humilis have reached their limit, or may yet become much taller.

pages 26–27: Cumulus mediocris rise beneath a layer of altostratus. Although substantial enough to darken the sky, as in this scene, mediocris clouds rarely produce rain.

left: Sunlight illuminates the sides of cumulus congestus clouds. The tallest of the cumulus clouds, the congestus can produce showers and darken the sky for observers beneath them. Cumulus congestus tend to form in clusters and, frequently, towering clouds of varying heights will congregate in what is known as a cumulus congestus complex.

CUMULONIMBUS

Given a suitably unstable atmosphere, the tops of cumulus clouds can grow to great heights before quite literally running out of steam, or by having arrived at the stabilizing influence of the lower stratosphere. When such a cloud top rises high enough for its drops to freeze, and produces heavy rainfall beneath, it is no longer merely a cumulus cloud, but a cumulonimbus. Unlike the comparatively miniscule cumulus, the scale of cumulonimbus requires an observer to be some distance away in order to take in its form in its entirety. Vast and imposing, the cumulonimbus evokes ominous images of blustering winds, heavy deluges, and bone-rattling thunder. Indeed, this may be exactly what the poor folks beneath the cloud are experiencing. Considering the violent weather associated with the cumulonimbus, it is curious to note that the great height of these clouds inspired the expression of bliss, "being on Cloud Nine"; of the ten cloud types enumerated in the original 1896 edition of the *International Cloud Atlas*, cumulonimbus was listed as number nine.

The lower portion of the cloud is composed of water-drops and has the familiar appearance of buoyant, cumuliform billows with clearly defined boundaries. This part of the towering cloud is often the source of summertime showers, which tend to be fleeting and limited in their horizontal span. In its upper reaches lie the more distinctive features of the cumulonimbus. Here, the cloud is a roiling mix of ice, snow, hail, and super-cooled water that has not yet frozen. At its top, where the cloud has lost buoyancy, the ice and snow is ejected from the core of the cloud

continued on page 34

and spreads to form the characteristic wide, flat top of an anvil cloud (see page 32). In contrast to the sharply defined edges of the lower cloud, the effect of ice crystals on reflected sunlight gives the top of cumulonimbus clouds a fuzzy and sometimes ridged appearance.

If clouds are the atmosphere expressing its moods, then the turmoil at the core of a cumulonimbus must be a display of its displeasure. Nevertheless, a lone cumulonimbus is not quite the foulest expression of atmospheric temperament; that accolade must be reserved for their congregation as tropical cyclones and tornado-spawning supercells (see pages 38–39). These formidable storms, formed from multiple cumulonimbus towers sharing a sprawling anvil cloud, span hundreds of miles and wreak devastation as they march across the landscape.

pages 32, 33, and 36–37: Cumulonimbus varieties (see page 172).

right: With the passage of a storm front, heavy rains and cumulonimbus may be accompanied by cumulus and stratocumulus, as in this mixed-cloud scene. To a mariner or other similarly exposed individual who has suffered through fierce storms, the arrival of broken clouds such as these would be a welcome development.

A lone lightning bolt strikes the ground beneath a supercell thunderstorm at sunset. Cloud-to-ground lightning like this is less common than bolts within the cloud itself. Both are sudden redistributions of electric charge, caused by all the violent collisions occurring among the cloud's raindrops and ice particles.

A shelf cloud sits beneath a cumulonimbus cloud on the leading edge of a supercell thunderstorm. Shelf clouds form as the outflow of air, dragged down by the motion of precipitation, spreads out at ground level like a carpet. The lower lip of this wall is sometimes called a "whale's mouth". Behind it, the cloud base rises into a yawning cavern.

left: Of all the signs of threatening weather offered to us by clouds, perhaps none is as ominous as the thunderstorm with a green hue. Anecdotally, such a scene is thought to be advance warning of a tornado; although this is not always the case, the filtering of sunlight that creates this optical effect can be an indication that heavy rain, high winds and lightning are not far off.

pages 42–43: Clouds typically provide a vivid indication of the upward motion of air. However, on rare occasions the base of clouds can reveal the downward motion of sinking air, as in these mamma (Latin for "breast") forming on the underside of an anvil cloud. Instability within a mature anvil leads to downdrafts that create these strange, globular, udder-like formations that protrude from the base.

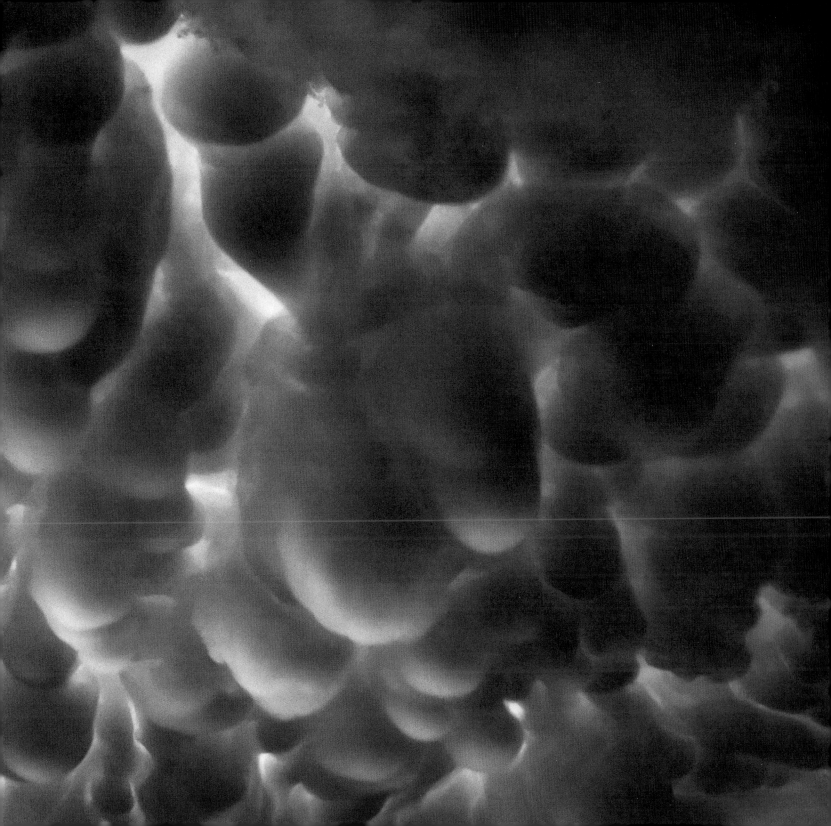

STRATUS

Stratus clouds are uniform, featureless blankets of dark grey cloud that form in the lower atmosphere. Just as the heaped appearance of the cumuliform clouds reflects their distinct pockets of rising air, so the layered appearance of stratus clouds reveals the gentle lifting and cooling of air over a broad area.

Stratus clouds are not known for their dynamic nature. "Stratus" is from the Latin "to spread out", and true to their name they tend towards sluggishness. In fact, their melancholy and even oppressive nature is likely to provoke similar traits in humans caught beneath them. Nevertheless, a stratus cloud can be stirred into action on occasion. Stratus at ground level, for example—which we know as fog—is often described as "rolling in", though this motion is best observed from above. Indeed, when a stationary stratus cloud forms as a result of the movement of humid air over a hill, the droplets moving with the wind and cloud can appear to flow like a stream over rocks. Under rare conditions, atmospheric waves will pass through a stratus cloud layer and transform a uniform layer of cloud into a heaving and rolling mass—a variety of stratus known as "stratus undulatus".

The key to the creation of a stratus cloud is the slow lifting of a large area of sky. Nature accomplishes this feat in two ways; one way is for the air mass to slide up the side of a hill or mountain. If the air is humid, the cooling induced by this lifting will produce a cloud; those formed in this fashion are called "orographic clouds". The second way a layer of air is raised is when a warm air mass collides with a cool air mass. Here, the cold air will effectively act as a wedge and force the warmer air upwards.

If the rather cheerless stratus cloud has one saving grace, it is that it does not generally produce rain. However, should you be caught in its incarnation as a low fog, you may still find yourself nearly as damp as if there had been a shower.

right: Draped over rolling hills . . . if a breeze is blowing, a stratus cloud like this may appear to be streaming over the crests, but in spite of such a fluid appearance, the entire layer of cloud will persist in the same location for hours, or as long as the winds continue to carry moist air into the stratus cloud layer.

pages 48–49: A few rays of sunlight find their way through gaps in an otherwise opaque stratus cloud. Rays such as these, which appear to spread as they emerge from a cloud hole, are called "crepuscular rays".

page 50: Valley fog (see page 172).

page 51: Stratus fractus and cumulus (see page 172).

right: An iridescent skyscape of altostratus and altocumulus disappears from view as a curtain of stratus is drawn beneath it.

STRATOCUMULUS

Each spring and early summer, a blanket of low clouds pushes in from the ocean and hangs over the famously sunny beaches of Southern California. Needless to say, this tidbit is usually left out of the tourist brochures, but the phenomenon is well known to the coastal residents, who refer to the season as the "May grey" and "June gloom". This persistent carpet of low clouds is a stratocumulus layer, and although it may be at most only a few thousand feet thick, it is typically thick enough to render a summer's day unexpectedly cold and dark for the unwitting beachgoers beneath it.

While clouds are relatively infrequent in the wider subtropics due to a general pattern of sinking air, Southern California's coastal location occasionally makes it a stratocumulus-blighted exception. The air that rises in the towering cumulonimbus clouds of the tropics subsides in the subtropics, acting to inhibit the growth of clouds there. However, along the eastern margins of the subtropical oceans, the moisture which rises from the sea interrupts this pattern. This moisture rises to feed bubbling cumulus clouds, while from above a stable, warm mass of sinking air effectively acts as a lid, trapping moisture close to the surface and forcing the cloud formations to spread out into a shallow layer rather than continuing to grow vertically.

The result is expansive decks of stratocumulus that can extend hundreds or even thousands of miles out to sea. Underneath these decks a cool, damp environment is created. Seen from above, however, these clouds have a far more pleasing effect, taking on the majestic appearance of endless fields of rolling hills. Rounded crests separated by shallow valleys mark the air's rising motion, lifting the top of the cloud.

This kind of subtropical stratocumulus, found over eastern ocean margins, is certainly the most persistent example of stratocumulus, but by no means the only one. Transient fields of stratocumulus can occur during fair-weather days in many locations, and can also occur adjacent to stormy weather. Nor do stratocumulus clouds have to exist as a solid deck; patches of broken stratocumulus are frequent over both land and sea.

A great irony of the stratocumulus is that, thanks to the warm, stable air above it, just beyond its deck there is concealed exactly the brilliantly blue, cloudless sky that makes for a perfect day at the beach.

Those unlucky enough to be caught underneath a stratocumulus deck will often be completely unaware of the crystal-clear blue sky just a few thousand feet above them. Stratocumulus clouds are frequently capped by a warm, stable mass of sinking air that suppresses the growth of tall clouds. This lid helps to maintain the stratocumulus as a shallow, overcast layer.

pages 58–59: "Hope" in the form of a clear sky and an orange sunset appears on the horizon from beneath a dark layer of stratocumulus clouds. This continuous deck gives way to broken stratocumulus and cumulus at its edge.

left: Decks of stratocumulus can seem to stretch as far as the eye can see. Nevertheless, holes in overcast decks, as well as sheets of broken stratocumulus, are not uncommon. And, as in this case, the seemingly interminable blanket of cloud does eventually pass, to be followed by a markedly brighter sky.

right: Thin, broken stratocumulus is a significant contrast to thick, overcast stratocumulus (shown on the following pages). As the sun begins to penetrate and warm the earth's surface below, it can hasten the erosion of this cloud deck.

pages 64 and 65: Stratocumulus (see page 172).

ALTOCUMULUS

The individual cells of altocumulus are known as cloudlets, and often appear as fleecy tufts or mounds, like smaller, more regimented versions of fair-weather cumulus clouds. Amassed in ranks at middle altitudes, they appear to, as Rupert Brooke eloquently put it, "ride the calm mid-heaven". Although it might look calm to the observer, in actuality this form of altocumulus can indicate instability rather than calmness up at cloud level.

"Mackerel sky, mackerel sky; never long wet, never long dry." This was the warning known to the sailors who referred to fields or rows of altocumulus cloudlets as a "mackerel sky" because of its resemblance to fish scales. The expression also reflects the fact that when an extensive layer of these tufts appears on a humid summer's morning, it is a good indication that thunderstorms will follow. This is particularly true of the altocumulus castellanus species, which displays tufts that have grown taller than they are wide.

When altocumulus cloudlets cover a large area of the sky, as is frequently the case when altocumulus is present, the layer is classified as stratiformis. Sometimes, the cloudlets are smooth, rounded masses—like buns on a baking tray. At other times they take the form of long, parallel rolls of cloud. The appearance of this species, known as undulatus, is often reminiscent of waves on the ocean. They can also be an indication that the winds at the top of the clouds are blowing in a different direction to the winds at their base—a condition known as wind shear.

One of the most dramatic species of altocumulus is altocumulus lenticularis. Occurring on the downwind (or leeward) side of mountains, it appears as a smooth individual disc or lozenge, and is often thought to look like an enormous flying saucer (see pages 78–79). As winds blow up and over a mountain, a wave forms in the layers of air above the mountain. Lenticularis clouds form in the crest of the wave if the air above the mountain is sufficiently humid.

Altocumulus clouds form at middle altitudes of approximately 6,000–25,000 feet (1,800–7,600m). Although often smaller than lower-altitude cumulus clouds, altocumulus tend to adopt the same heaped appearance produced by rising plumes of buoyant air. Within the unstable cloud layer, it is the rising air that produces altocumulus cloudlets and the sinking air that accounts for the voids between them.

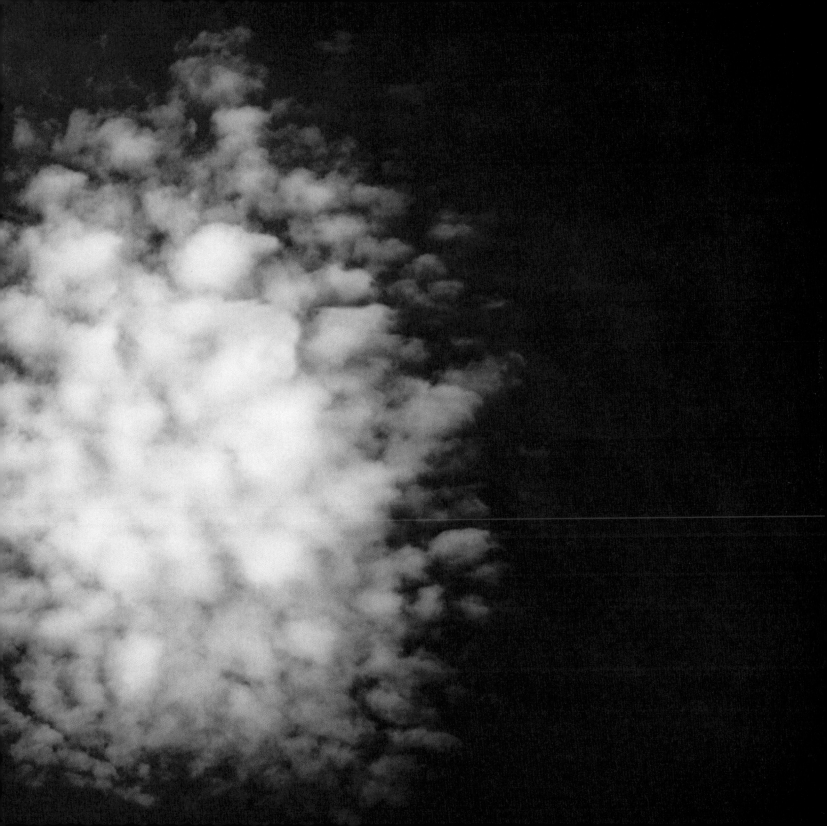

The ragged appearance of altocumulus floccus suggests that there was previously a thicker layer of altocumulus that is now dissipating.

This altocumulus layer is similar to stratocumulus, but it is at a higher altitude. The small gaps between cloudlets indicate higher humidity.

A high, snow-toped mountain peak is capped by an orographic cloud, a cloud that develops in response to the forced lifting of air by topographic features.

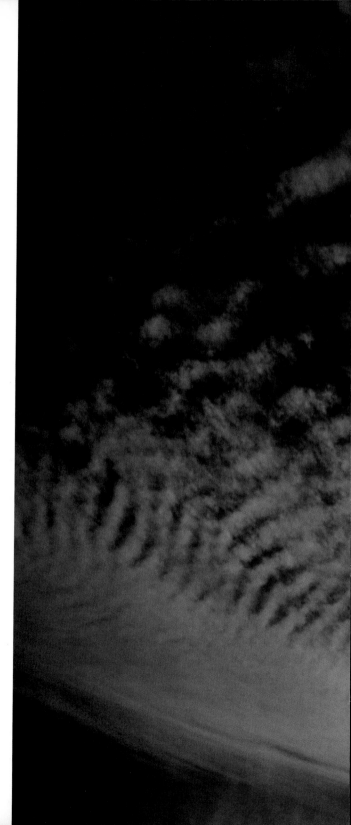

right: The layered structure of this cloud indicates that it is the stratiformis species of altocumulus, but it also has additional features. Altocumulus clouds come in seven varieties, and this formation exhibits characteristics of at least two of those: undulatus for the undulations generated by waves in the air, and translucidus— denoting that the outline of the sun is visible through the cloud.

pages 76–77: Collections of altocumulus clouds have been likened to flocks of sheep, migrating herds of buffalo, and even spacecraft. What forms are hidden in the shapes of these unusual patches of altocumulus?

Not to be mistaken for an Unidentified Flying Object (UFO), an altocumulus lenticularis cloud takes on a distinctive lens shape. Lenticularis clouds are orographic clouds that form in the crests of waves in the air. In this striking example, layers of air well above a mountain have been disturbed by the winds blowing up and over the mountain-top, leading to a leaning stack of clouds

ALTOSTRATUS

Altostratus clouds are featureless layers of cloud in the middle altitudes. In their most common state, when they cover the sky with a uniform greyness, there is little to appreciate. However, on rare occasions altostratus can put on a show. When an altostratus cloud is on the eastern horizon, and the sun is setting to the west, the uniform grey on the underside of the cloud becomes a vibrant display—one that Henry David Thoreau likened to "the stain of some berries crushed along the edge of the sky".

Altostratus can be difficult to distinguish from the lower-level stratus clouds. The principal distinction between them is altitude, although the German philosopher and poet Goethe also observed that "the upper clouds were like streaked wool, the lower heavy". The differing quality of the low and high clouds can be attributed to differences in composition: while stratus clouds are composed entirely of water droplets, the colder temperatures at higher altitudes mean that altostratus are frequently composed of both water drops and ice crystals.

Altostratus clouds exist in several varieties, all grey. A thin layer of altostratus translucidus is so-called because it allows the disc of the sun to be seen from below, while the unfortunately much more common variety altostratus opacus obscures the sun. The only variety exhibiting any texture is the undulatus, organized into rows by waves in the cloud layer.

continued on page 85

As with lower stratus clouds, altostratus occurs when an entire layer of air is lifted and cooled. In the middle altitudes this is usually as a result of a passing warm front. A warm front signifies the location where a layer of warm air collides with a layer of cool air. The warm air will lift above the cooler air and altostratus will result if the warm layer is humid. In weather lore, all middle- or high-level clouds tend to be regarded as harbingers of bad weather. Of these, however, a thickening layer of altostratus is the most dependably foreboding, particularly if it has been preceded by a progressively thickening and lowering layer of cirrus and cirrostratus.

pages 82–83: Altostratus (see page 173).

left: "Red skies at night, sailors delight. Red skies in the morning, sailors warning." This is perhaps the best-known adage of weather lore. Altostratus in the western sky will be illuminated by a rising sun in the eastern sky, providing a warning of approaching weather.

At the tail-end of twilight the vibrant colors fade from the lower surfaces of altostratus and the cloud layer resumes its customary, anonymous, monochromatic appearance.

When high winds provoke instability in the cloud layer, altostratus may take on a wavy pattern to become altostratus undulatus. This altostratus layer has been stretched into a ribbed pattern by winds blowing in different directions above and below the cloud.

Altostratus undulatus is the only altostratus variety to exhibit a texture. Undulatus can be produced by waves in the winds above, below or at the cloud layer. Such undulating cloud decks are also known as "billow clouds" or "wave clouds".

In a rare moment of dramatic transformation, an altostratus cloud has been bowed by the deflection of a strong wind over a mountain-top.

NIMBOSTRATUS

If we are to be truthful, it must be admitted that our regard for clouds is inconsistent. Clouds spark the imaginations of children, just as they have inspired great poets and painters, yet alongside tributes to the brilliant shades of red and orange painting the underside of a stratus cloud at sunset, there are references to poor judgment "clouded" by incomplete understanding or accusations that a disappointment has "rained on our parade". Of all the expansive, brooding clouds that inspire the gloomy phrases, certainly the most deserving of such usage is the nimbostratus, which typically develops from a thickening layer of altostratus. As the cloud base lowers, it becomes a featureless, grey expanse. Now opaque, the disk of the sun is no longer visible, the scene becomes quite dark, and precipitation begins to fall.

The first widely adopted classification of clouds was prepared by Luke Howard in 1803. He lumped all of the raining clouds into a single cloud type: the nimbus (Latin for "cloud"). Not until the early twentieth century did the meteorological community establish the nimbostratus type, thereby drawing a distinction between this and the more turbulent cumulonimbus. In contrast to the heavy, but usually small-scale, downpours resulting from cumulonimbus, the rain from nimbostratus will cover a larger area and last much longer. Nimbostratus form from the widespread lifting that occurs along a front as a warm air mass is forced up the sloping surface of a cool air mass. Rain is not the only product of nimbostratus; snow and sleet will also fall from these clouds in colder climates. Although Howard's classification of rain-clouds was later changed, one of his contributions to the understanding of clouds was to refer to their types as "modifications", in part to draw attention to the fact that clouds will frequently change from one type to another. The nimbostratus is an example of just such a modification. In the mid-latitudes, high cirrus and cirrostratus mark the first indication of an approaching storm front. Many hours or a day later, as the front draws near, that same layer of cloud has thickened and lowered to become a dense nimbostratus.

Nobody looks forward to seeing a nimbostratus cloud, but it is important to note that were it not for rain-clouds we would not have rainbows. And herein lies the dichotomous nature of this cloud type, so perfectly noted by Luke Howard himself: "the nimbus . . . superbly decorated by its attendant the rainbow; which can only be seen in perfection when backed by the widely extended uniform gloom of this modification."

Sunlight illuminates only a small portion of this nimbostratus. The broad cover and thick nature of the cloud type often renders both it and the environment below dark and dreary.

left: This cloud is exhibiting two of the nimbostratus' characteristics: the first is falling rain, the second is the apparent lack of any other features. Perhaps Luke Howard had a cloud like this in mind when he described nimbus clouds as "widely extended uniform gloom".

pages 98–99: Like hangers-on, "pannus" clouds are small, dark shreds of cloud that tend to tag along with nimbostratus and other raining clouds. Also known more evocatively as "scud", these fragments of cloud form low in the atmosphere where evaporating rain has left the air saturated.

If one were to observe the formation of nimbostratus clouds over the course of a day, they would be seen to result from a slow and steady progression of high, thin cirrus clouds descending and becoming darker and more menacing in appearance.

CIRRUS

Often thin and delicate, and rarely coincident with bad weather, cirrus clouds may frequently go unnoticed. There is much to be appreciated, however, by those who take note of their fibrous tendrils and varied forms. The loftiest of all clouds, cirrus reside at altitudes of between 24,000 and 45,000 feet (7,300–13,700m), and are shaped by the cold temperatures and high winds at these heights. Cirrus are composed of ice crystals, which once formed, are whipped by the winds into strand-like streaks. In fact, the name "cirrus" is derived from the Latin term for a lock of hair.

Although these clouds are usually observed during fair weather, mariners have long known that cirrus can be a harbinger of bad weather. In the mid-latitudes, the first indication of an approaching storm front is cirrus developing across the western sky. Severe thunderstorms can also actively generate cirrus that can extend far from the area of disturbed air. However, the presence of cirrus overhead is an imperfect indicator of future weather. Prior to the sophisticated weather forecasts of the modern age, anyone whose livelihood or wellbeing was tied to the weather would have had to pay careful attention to the development of cirrus clouds in order to determine whether they were truly indicating deteriorating weather, or only a remnant of a long-dissipated storm.

Cirrus come in five forms, identified by the appearance of their characteristic streaks. Cirrus fibratus are composed of long parallel tendrils which do not culminate in a tuft of cloud. When these tendrils take on a curved or hooked appearance, they are called cirrus uncinus, known to mariners for generations as "mare's tails"—its curved tails are formed when ice crystals fall through layers of the atmosphere which have varying wind speeds. The thickest cirrus variety is cirrus spissatus, a grey patch of high ice cloud that is often the last remnant of a tall thunderstorm. Cirrus floccus consists of individual tufts of cloud, usually with a trail of long tendrils emanating from each tuft. Cirrus castellanus is the name for individual high clouds in the form of turrets with a common level of cloud base. Although each cirrus variety is distinctive, it is common to see several of them grouped together to form a single cloudscape.

Nautical lore tells us that "mares' tails and mackerel scales make lofty ships carry low sails"—a scene such as this, with a combination of the characteristic hooked appearance of cirrus uncinus's mares' tails accompanied by dappled cirrocumulus above them, might be the earliest indication of an approaching storm. When the high winds of a storm are blowing across the deck of a tall ship, the sailors will have to reduce their sails in order to remain upright in the water.

Not all cirrus clouds are produced by storms. Cirrus forming during fair weather merely indicates an instability over a limited area. Pictured during the early stages of forming, these cirrus clouds appear as clumps of rising air, generating small ice crystals that are only beginning to grow into the silky threads of cirrus.

Cirrus uncinus is the formal name for cirrus tendrils that appear as hooks or commas. Nicknamed "mares' tails" by mariners, they are also known as "fall streaks". Here, ice crystals are seen dropping out of a fast-moving layer of clouds, leaving curved trails in their wake.

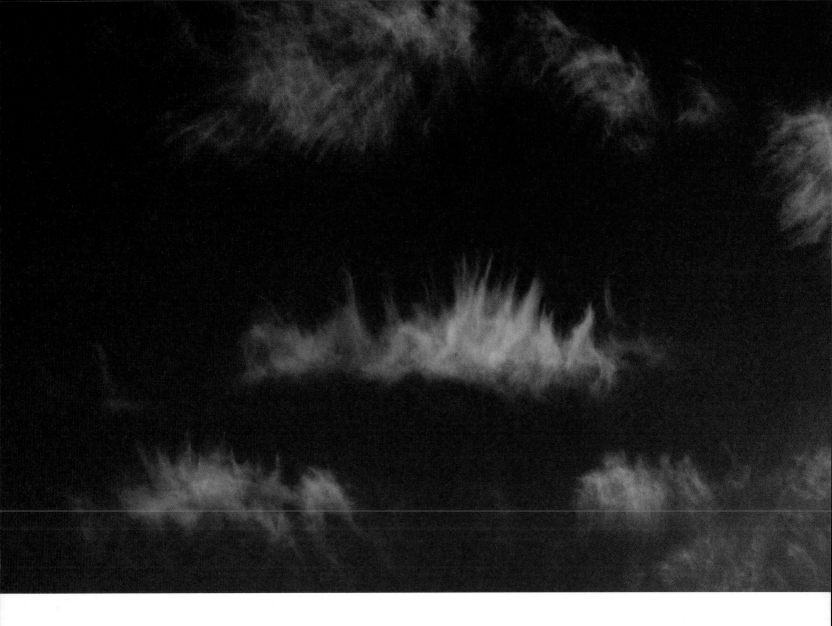

When ice crystals linger for long periods without evaporating the results are long trails of fibrous cirrus. If, as in this scene, the air is a bit less humid, ice crystals settle only briefly from their point of origin before turning back into invisible vapor.

pages 110–111: Cirrus spissatus is the thickest and most turbulent-looking of the cirrus species, suggesting that stormy weather is not far away. Tall thunderstorms loft large amounts of moisture to their upper reaches, which flows away from the storm. Thick cirrus commonly lines the edges of this outflow.

left: The higher the altitude, the faster the winds blow. At the height of cirrus clouds, ice crystals are whipped about by fierce winds. It is easy, then, to imagine how lingering cirrus may quickly spread across the sky. Cirrus filaments in this cloud deck may eventually become a finely striated layer of cirrostratus fibratus.

CIRROCUMULUS

Cirrocumulus clouds are composed of groups of cloudlets arranged in a high, thin layer. The unassuming cloudlets are arranged either in patches of closely collected cloud mounds or organized into rows—both resulting in the creation of majestic ranks of cirrocumulus that stretch across the sky. Unfortunately, this striking formation is often fleeting, perfectly illustrating the transient nature of clouds. The cirrocumulus cloudlets may ultimately blend together and stretch out into a fibrous layer of cirrostratus, or perhaps separate into distinct locks of cirrus.

Cirrocumulus and altocumulus are both composed of these small cloudlets, and the two can be difficult to distinguish, though cirrocumulus cloudlets are typically smaller than altocumulus ones. Furthermore, individual altocumulus cloudlets may be partially in the shadow of a neighboring cloudlet; in contrast, cirrocumulus cloudlets never cast shadows upon each other.

From time to time, near where the sun or moon is shining through a layer of cloud, the fringes of the cloudlets can display a vivid array of colors. On the rarest of occasions, concentric rings in pastel shades of green, yellow, orange, and pink will appear to surround the sun or moon in an optical effect known as a corona. The colors result from the slight redirection of light rays as they are scattered among the cloud drops or ice crystals on their way through the cloud layer. Iridescent clouds, which are simply fragments of an incomplete corona, are a similarly pleasing sight and considerably more common than a full corona. Apart from the fringes of thicker clouds, thin clouds with small, uniformly sized drops or crystals are most likely to produce iridescence and coronas, which is why cirrocumulus and altostratus are the best candidates for creating these effects.

A uniform layer of cirrostratus cloud gives way to cirrocumulus undulatus at its edges. High cirroform clouds often share the same skyscapes, and, as shown here, transform from one type to another.

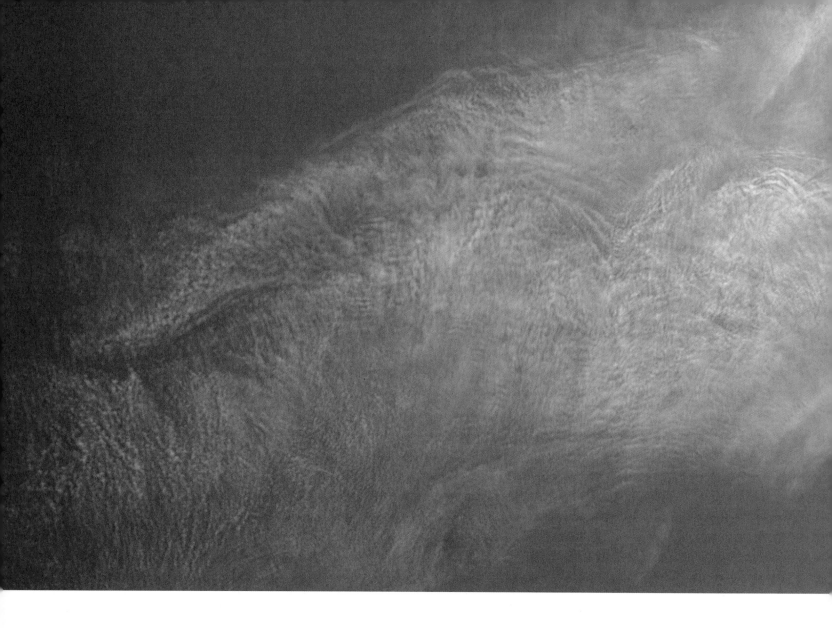

Like a perforated cloud, the unusual cirrocumulus lacunosus exhibits small round gaps fringed with tiny cirrocumulus cloudlets. The resulting cloud layer takes on the appearance of a net or honeycomb.

High winds are a frequent companion of high clouds. This line of
cirrus clouds has generated ice crystals, which are being swept along
by the winds and are partially obscuring a layer of cirrocumulus.

pages 120–121: This cloud layer appears to be making the transition from cirrocumulus to cirrus. This is suggested by the separation between the cloudlets and the production of virga, also known as "fall streaks", which appear as the tails of ice crystals descending from isolated cloudlets.

right: This cirrocumulus cloud exhibits a variation of "mackerel sky"—a layer officially classified as cirrocumulus stratiformis undulatus, with rows of cirrocumulus cloudlets arranged in a ribbed pattern that is named for its similarity to the scales found on the King Mackerel fish.

pages 124–125: With its features softened and its rows of cirrocumulus remaining only faintly, this "mackerel sky" is in the process of dissipating and will soon have vanished from the sky.

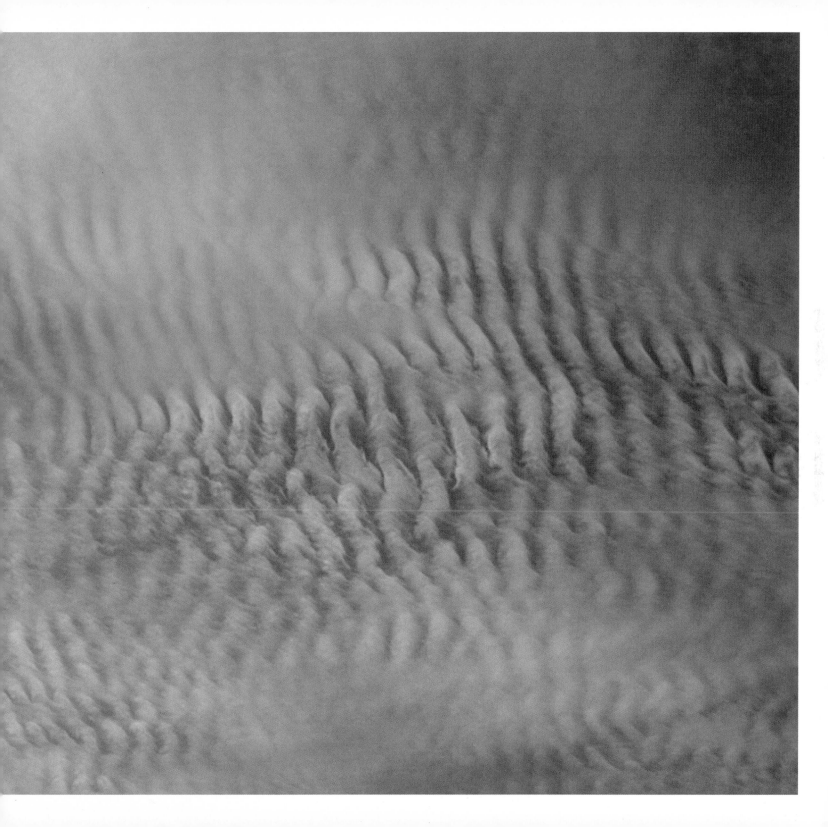

CIRROSTRATUS

The highest of the cirroform layer clouds is cirrostratus. Drawn like a delicate curtain across large portions of the sky, cirrostratus clouds are thin, sometimes barely perceptible, layers of ice crystals. They are nearly always transparent, allowing the disk of the sun to be clearly visible through the cloud layer.

The transparency of cirrostratus, together with the tendency for its ice crystals to bend the rays of the sun, creates the most distinctive feature of this cloud type: the halo. Described by the poet Shelley as a "girdle of pearl", the halo appears as a bright ring around the sun or a luminous full moon. If the ice crystals take the form of tiny, six-sided flat plates, the halo may be accompanied by one or two "false" suns, bright spots of focused sunlight known as "sun dogs" that appear at a distance to the side of the actual sun.

These optical phenomena bring attention to a cloud type that might otherwise be overlooked. The hazy quality of a type of cirrostratus that isn't accompanied by these visual effects is reflected in its name, cirrostratus nebulosus. One of the two recognized species of cirrostratus, it is so-called due to its lack of identifiable textures or features and in its thinnest incarnation, cirrostratus nebulosus may be invisible to the eye. The second is cirrostratus fibratus, where the layer is finely striated as the tendrils of ordinary cirrus combine and spread to form a single deck. Either species may be observed in undulatus form, where the cirrostratus layer appears wavy, or the duplicatus variety, where two distinct layers are present.

Alerted by the distinctive presence of a radiant halo, early sky-watchers took note when the otherwise understated cirrostratus spread across the sky. Like the appearance of mackerel scales or a red sky at night, cirrostratus halos are understood to be a warning of bad weather to come: "when a halo rings the moon or sun, rains approaching on the run."

It is not uncommon for cirrostratus to form broad layers of cloud that cover much of the sky. Being at high altitudes means their surfaces are perfectly positioned to generate vivid sunrises and sunsets. As with their mid-level relatives altostratus, a thickening layer of cirrostratus often indicates approaching "weather".

This layer of cirrostratus thins as it stretches from the horizon to the foreground. In fact, it is so thin overhead that were it not colored orange and pink by the setting of the sun, it might not be seen.

The divergent motions of these two layers of cirrostratus suggest that the winds at each layer are blowing in different directions. The simultaneous occurrence of two distinct layers of any type of cloud is classified as the duplicatus variety. Waves generated by instability in the winds in the upper layer will yield cirrostratus undulatus.

OTHER CLOUDS & PHENOMENA

The ten principal cloud types, from cumulus to cirrus, all exist within the troposphere, the lowest layer of the atmosphere that extends from the earth's surface to about 8 miles high (13km). A few types of cloud, however, can be found forming at higher altitudes. Rarely spotted, even by dedicated sky-watchers, these clouds tend to be thin and located near the north and south poles. Two such cloud types worthy of mention are polar stratospheric and noctilucent clouds.

Polar stratospheric clouds are found, as the name suggests, near the poles and in the stratosphere—the atmospheric layer directly above the troposphere. They are also known as nacreous clouds or "mother-of-pearl clouds" due to their lustrous, opalescent appearance. The vivid display of pastel colors by nacreous clouds, called iridescence, results from the scattering of sunlight as it passes through cloud particles—the same process that produces coronas.

Extremely thin and delicate clouds also form in the mesosphere, the atmospheric layer directly above the stratosphere. Light blue and fibrous in appearance, these clouds are known as noctilucent (meaning "night-shining") clouds, as they can only be observed at twilight, when the sun is behind the horizon. Presumably, they are present at other times, but are so thin as to be imperceptible except under the right conditions. Because of the difficulty in observing noctilucent clouds, their mechanisms of formation remain largely a mystery.

Rainbows are the best known of all cloud optical effects. They are only visible when viewing a rain-cloud with the sun behind you, because rays from the sun pass through the raindrops before being reflected, as a rainbow, back toward the observer. As the light enters the cloud drops, they act as tiny prisms, in which the rays of different wavelengths, or colors, are bent by different amounts. The result is the familiar band of colors, neatly ordered from red to violet. On rarer occasions, two, or even three, bows may be observed. Fainter than the primary bow, these supplementary arcs occur when light reflects two or three times within a raindrop before exiting.

If the old saying is true, no less than four pots of gold may be found at the endpoints of this double rainbow. Rainbows are observed when sunlight reflects off raindrops. The secondary bow, which is only infrequently observed, results from sunlight reflecting twice off the surface of the raindrops. Note that the order of the colors is reversed in the secondary bow.

pages 138–139: A neon green aurora arches over layers of faint noctilucent clouds on the horizon. Auroras are caused when particles emitted by the sun are focused on the poles by Earth's magnetic field and collide with atmospheric particles as they pass through the atmosphere. Auroras, noctilucent clouds, and nacreous clouds are just some of the unique atmospheric phenomena observed near the north and south poles.

right: Nacreous clouds, informally known as mother-of-pearl clouds, are observed only rarely and almost always at high latitudes. Broadly undulating and vividly iridescent, these distinctive clouds are an example of polar stratospheric clouds; clouds that form at an altitude of 12–19 miles (20–30km), well above the highest cirrus.

left: Mysterious, and even slightly spooky, the light blue fibres of noctilucent clouds appear to onlookers only at twilight when the sun is over the horizon, but these clouds, which are located in the mesosphere more than twice as high as the highest cirrus clouds, remain illuminated by sunlight.

pages 144–145: Dust, smoke, pollution, and sea salt are common substances found drifting about as particles in the atmosphere. Here, an accumulation of particles has swollen by collecting molecules of water from the humid air that surrounds small cumulus clouds. The resulting haze fades the blue of the sky and reduces visibility.

CLOUDS:
A VISUAL GLOSSARY

EARTH'S ATMOSPHERE

The Earth is wrapped in a thin shell of gases, held captive to the planet by the pull of gravity. Although unable to escape into space, the air is by no means locked in place. To the contrary, Earth's atmosphere is a dynamic place—we can experience its turbulent nature as the wind on our face or the rustling of treetops on a blustery day. Were it not for the presence of clouds, however, we would have little knowledge of the activity occurring above the treetops. The atmosphere is always in motion, yet the molecules of nitrogen, oxygen, water vapor, and other trace gases that compose the air we breathe, are invisible to our eyes, so we see little evidence of this motion on a clear day. In addition to being elements of aesthetic wonder, clouds provide us with a window through which we can observe these otherwise hidden dynamics of the air above us.

The popular astronomer Carl Sagan described the Earth viewed from billions of miles out in space as a "pale blue dot". Our planet appears blue because 70 percent of its surface is covered with oceans. Clouds reflect approximately 15 percent of the sunlight striking the Earth and their brilliant white formations help explain why the Earth appears pale. Forever shifting and evolving, clouds change hour by hour. Nevertheless, even a single snapshot from a weather satellite reveals some of the global patterns of cloud cover. In the tropics, for example, a narrow band of cloud cover girdles the planet along the Intertropical Convergence Zone, where the heat from the tropical sun is converted into violent cumulonimbus clouds rising to 50,000ft (16,000m) above the Earth's surface. Further from the equator in the calmer subtropics, the rising air from the tropics sinks back down, creating frequent high pressure and relatively clear skies. However, along the subtropical west coasts of the continents, cold ocean currents from higher latitudes combine with humid maritime air to produce persistent decks of low stratocumulus clouds that may linger for days at a time. Further poleward are the strong westerly winds known as the Roaring Forties and Furious Fifties; named for their location between 40

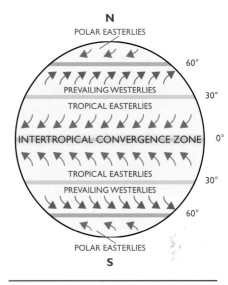

BASIC WIND PATTERNS OF GLOBAL ATMOSPHERIC CIRCULATION

Global wind patterns include the steady trade winds from the east (in the tropics) which converge near the equator, creating a band of cumulonimbus clouds. Between 40 and 60 degrees latitudes, strong westerly winds known as the Roaring Forties accompany mid-latitude cyclonic storms in their march eastward around the globe.

OPPOSITE: Regional patterns of cloud formation are apparent when the Earth is viewed from space. The intertropical convergence zone, near the equator, is lined with cumulonimbus cloud systems. Further south, along the subtropical coast of Peru, an extended deck of stratocumulus clouds stretches out to the central Pacific Ocean. Further south, a mid-latitude cyclone comes ashore along the southern Chilean coast.

and 60 degrees from the equator in both the northern and southern hemispheres. These regions are the domain of the mid-latitude cyclones—frontal cloud systems that follow one another in a slow, eastward movement around the planet.

HEAT TRANSFERENCE

Although it is undeniably complex and chaotic, the Earth's weather helps to perform the vital function of redistributing heat. The atmosphere is largely transparent to sunlight. On a clear day, much of the sunlight reaching Earth will pass through the atmosphere and heat the planet's surface, although the effect will be reduced if the surface is lightly colored and reflective, as with deserts or snow, or if the skies are cloudy. The Earth's lower atmosphere is therefore not unlike a pot of water on a stove—it is a fluid being heated from the bottom, creating an unstable situation. But the stirring of the atmosphere is only partly explained by heating from below. The heat from the sun is not distributed evenly; the tropics are warm because the sun is high overhead all the year round, while at the much colder North and South poles the sun is much lower in the sky and the solar

energy absorbed there is much reduced. The excess heat absorbed in the tropics has to find its way toward the poles, yet it is accumulating much too fast to merely diffuse, and must be transferred by the winds. So the global weather patterns we observe are the expression of the atmosphere's adjustment to the instabilities constantly generated by the heating power of the sun. The churning and roiling of the weather helps transport heat upward and poleward away from the tropics.

THE "SPHERES"

The atmosphere can be divided into five layers determined from the vertical

ABOVE: A photograph taken from the International Space Station reveals the layers of the atmosphere. The transition from orange to blue colors occurs at the troposphere and stratosphere boundary.

structure of air temperature. The lowest, the troposphere, extends from the Earth's surface to about 8 miles (13km) altitude. Above the troposphere is the stratosphere which reaches to about 30 miles (50km). The mesosphere extends to about 50 miles (80km), the thermosphere to about 300 miles (500km), and the exosphere to perhaps 6,000 miles (10,000km), which is beyond the nominal boundary of outer space. The weather we experience, and the vast majority of clouds, occur within the troposphere. Small amounts of cloud are also observed in the stratosphere and mesosphere. Within the troposphere, temperature decreases with altitude, creating the potentially unstable conditions which help to maintain weather systems and their attendant cloud systems.

The stratosphere is home to elevated concentrations of ozone, a chemical that is efficient at absorbing ultraviolet light and benefits life below because it reduces the transmission to the Earth's surface of ultraviolet radiation, which is toxic to many biological structures. This absorbed ultraviolet light heats the stratosphere so that it is warmer than the top of the troposphere below. When the temperature increases with altitude, as in the stratosphere, vertical motions are inhibited. If, for example, a pocket of air were displaced upward in a gust of wind, it would move up into a region of warmer air. We all know the rule that warm air rises. This is because warm air has a lower density than cool air (see discussion of buoyancy below). In the case of our displaced parcel of air in the stratosphere, however, it is cooler and more dense than its new surroundings, and therefore drops back down to where it started. Hence, small displacements of air are inhibited by the temperature structure of the stratosphere and layers within it tend to be stable. The prefix "strato-" is derived from the Latin for "spread out", in recognition of the stable layered structure of the stratosphere. The weak motions in the stratosphere mean that clouds here are infrequent. However, the tops of very strong thunderstorms can penetrate the tropopause (the boundary between the troposphere and stratosphere) and deposit small amounts of cloud into the lower stratosphere, which then spread out into thin layers. Thin stratospheric layer clouds also form at the poles.

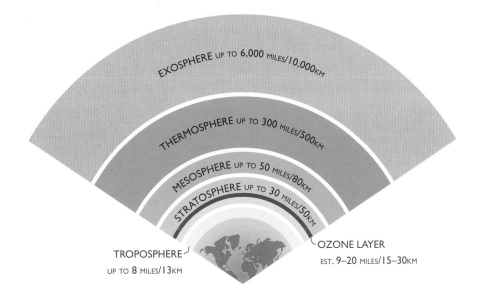

EXOSPHERE UP TO 6,000 MILES/10,000KM

THERMOSPHERE UP TO 300 MILES/500KM

MESOSPHERE UP TO 50 MILES/80KM

STRATOSPHERE UP TO 30 MILES/50KM

TROPOSPHERE
UP TO 8 MILES/13KM

OZONE LAYER
EST. 9–20 MILES/15–30KM

LAYERS OF THE ATMOSPHERE

There are five principal layers of the atmosphere. The clouds and weather we experience occur almost entirely within the lowest layer, the troposphere, where temperature decreases with altitude. Thin layer clouds also occur in the stratosphere and mesosphere, mainly near the poles.

THE WATER CYCLE

Water passes through clouds as part of a complex cycle where it is continuously shifted from the ocean, to the atmosphere, to the land, and back to the ocean. Important reservoirs of water on land include lakes, rivers, glaciers, and ground water.

Temperature decreases with height again in the mesosphere, and vertical motion here does occasionally lead to the formation of very thin clouds. Generally these clouds are only visible at dusk or dawn when the sun appears to be below the horizon to an observer on the Earth's surface but above the horizon from the perspective of clouds in the mesosphere.

HOW CLOUDS FORM

A cloud is merely a collection of water drops and ice crystals, but where do those come

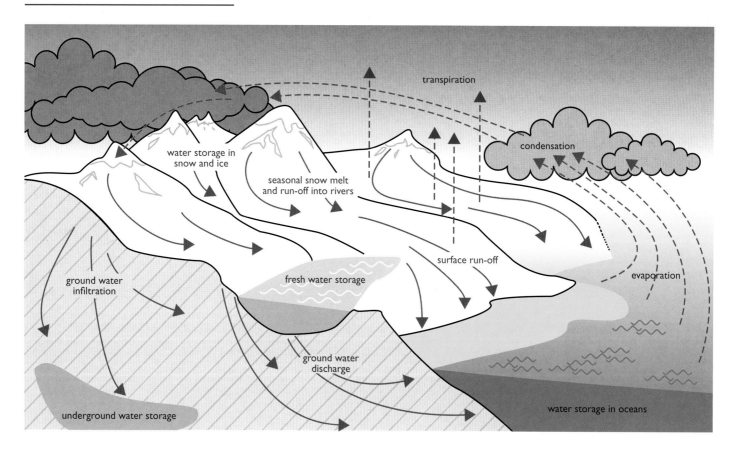

from? Within any mass of air there are molecules of water vapor jostling among the other molecules. Water molecules have a tendency to bond with one another to form a liquid, but only if they are moving slowly enough for the weak bond between the molecules to hold. The temperature of air is a measure of the average speed of the molecules, so a lower temperature means slower molecules and improved chances that colliding vapor molecules might bond together. The process of vapor molecules accumulating into liquid water is known as condensation. Condensation of vapor molecules into cloud droplets (a term used here interchangeably with "drop") begins when a small amount of water vapor accumulates on particles, such as dust or sea salt, that are suspended in an air mass. As that mass cools, it will reach a point where more of the vapor molecules have slowed enough to maintain bonds with water molecules that have already accumulated on particles, This temperature is known as the dew point temperature, and marks the point where wet particles begin to grow into cloud droplets. Further cooling of the air will promote additional condensation and the cloud droplets will grow.

Why is the air cooling? Clouds form most frequently as a result of rising air, expanding as it ascends. Air becomes thinner with altitude, as anyone who has climbed to the top of a tall mountain can attest, which means that for a given volume there are fewer molecules of air the higher the altitude. The higher a climber goes up a mountain, the less air there is above pressing down on him or her; likewise, there is less air pressing the molecules of air together in a cloud as they rise. As the air rises within a cloud, the molecules of gas slow down and move farther apart. The slowing of the molecules promotes further condensation of cloud droplets, as noted above. Although rising air is responsible for the formation of most clouds, this is not always so—one exception to this rule is fogs. Foggy air remains at ground level, rather than rising, and cools by another process, such as making contact with a cooling surface below.

Despite the many clouds illustrated in this book, it is important to appreciate that individual cloud types rarely occur in isolation. Even in fair weather it is not uncommon to see a patch of small cumulus clouds with a separate layer of altostratus or cirrus thousands of feet above. And when there are storms—the most prodigious generators of clouds—a variety of cloud types will be found in patterns that reflect the structure of the storm itself. Understanding the sequence of cloud types in a passing storm can even allow an observer to make simple predictions. Long before there were satellites, radars and television meteorologists, it was common practice among generations of mariners and farmers to observe and interpret the condition of the sky in order to glean information about forthcoming weather.

This glossary explores the relationship between meteorology and the types and arrangements of clouds associated with typical weather patterns and storms. Beginning with a discussion of fair-weather clouds, which are not associated with violent conditions and offer a situation most likely to yield clouds of a single classification, there follows a discussion of various storm types and the organization of the forms of cloud that results from them. The glossary closes with a discussion of a few of the less common types of clouds as well as a review of some of the optical effects that are associated with clouds, such as rainbows and halos.

CLOUD TYPES AND METEOROLOGY

FAIR-WEATHER CLOUDS

The fair-weather clouds are of the ubiquitous sort and as such may often go unnoticed. They can take several forms, reside in any location, and will occur at a wide range of altitudes, from low to very high. What these clouds share in common is that they are not associated with threatening weather. Be they small, puffy cumulus, a mackerel blanket of altocumulus, or tendrils of thin cirrus, fair-weather clouds give the sky some character without affecting human activities by inflicting rain or violent weather.

CUMULUS

Small cumulus clouds occur frequently in many climates—the tropical oceans near the equator are frequently decorated with fields of small clouds just above the ocean's surface, which are known as tradewind cumulus. The warming of a continental land mass on a sunny morning will heat the bottom of the atmosphere, which may lead to the growth of fair-weather cumulus in the afternoon.

Within cumulus clouds are parcels of air that have become buoyant. The density of these parcels is less than that of the surrounding air, meaning they are lighter, and this yields a force pushing them upward, known as "buoyancy forcing". This process is taking place whenever we see boiling water churning on a stove or a hot air balloon floating in the distance. In either case, a parcel of the fluid (for example, the boiling water in a rising plume or heated air inside the hot air balloon) is warmer and less dense than its surroundings and creates an upward force. In a similar fashion, cumulus clouds form where small parcels of warm air have been displaced upward into a layer of cooler air, such as by a random "bump" from a wind gust, and continue to rise because they are now warmer and less dense than the surrounding air.

Fair-weather cumulus, such as the cumulus humilis or cumulus mediocris species, are, by definition, shallow and not precipitating. The growth of these clouds is frequently capped by a stable layer above the cloud which is warmer than the cloud layer. The base of this stable layer, which coincides with the tops of the cumulus clouds, is called a

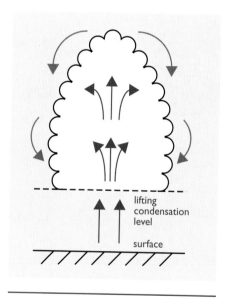

BASIC CUMULUS CLOUD

Inside cumulus clouds are buoyant plumes of rising air. The base of the cloud is formed at the lifting condensation level where rising air has cooled to the dew point temperature and water vapor begins to condense into cloud drops. The rising air below the cloud is replaced by subsiding air outside the cloud.

ABOVE: Typically small fair-weather cumulus clouds of the type that commonly grow over coastal hills in response to the warming of the land by the sun during the day.

temperature inversion because temperature is increasing with height there, in contrast to the typical condition where temperature decreases with height. Rising air within the cloud that reaches the temperature inversion is no longer buoyant because the surrounding air is warmer, and the loss of buoyancy prevents the cloud from growing taller.

Fair-weather cumulus are so named because they are associated with fine weather conditions. Indeed, the temperature inversion that limits the growth of these clouds forms in response to general patterns of fair weather. In the middle of storms, a large amount of air is moving upward, giving rise to thick clouds (see discussion of storm systems below). However, what goes up must come down, as the saying goes, and the regions of fair weather, such as those appearing as high-pressure regions between storms on a weather map are characterized by the widespread slow descent of air in the middle of the troposphere. It is exactly this tendency for the air to descend that prevents cumulus clouds from growing tall. While the sunny skies may help to promote a layer of warm, moist and unstable air near the Earth's surface, the descending air above establishes the temperature inversion within the first few thousand feet above the surface, limiting cloud growth.

STRATOCUMULUS

Small stratocumulus clouds also form under circumstances similar to those producing fair-weather cumulus. Residents of coastal areas along the western edges of the continents in the subtropical band of latitudes (locations such as Southern California, Peru, and the Azores) will be familiar with the persistent decks of stratocumulus clouds that are often found in these locations. Although rarely extending far inland, subtropical stratocumulus cloud decks frequently extend hundreds of miles out to sea. The subtropics are meteorologically characterized by the widespread slow descent of the air that rises in the tropical storms that form in the intertropical convergence zone (see diagram, page 149).

Over the subtropical oceans, steady evaporation from the surface of the water

155

leaves the lowest layer of the atmosphere
(the so-called boundary layer, extending
from the Earth's surface to no more than
4,000ft [1,000m]) very humid. As in the
case of fair-weather cumulus, a small
displacement of air upward provides an
opportunity for some of that moisture to
condense. However, over the ocean the
vertical motions in cumulus clouds tend
to be gentler. The high humidity near the

surface, in combination with the strong
subsiding air above, acts to trap all of
the moisture in a narrow layer, typically
no more than several hundred meters
thick, which remains saturated and filled
with cloud. Viewed from below, these
clouds may not reveal much structure
and therefore may be hard to distinguish
from stratus clouds. From above, however, the buoyant overturning of the cloud
layer is usually evident in an expanse of
undulating cloud. Occasionally these
clouds may drizzle, allowing some of the
trapped moisture to return to the ocean.
After a portion of a stratocumulus deck
drizzles, that part of the cloud deck may

appear broken with dregs of dissipating
cloud, small detached cumulus clouds, or
even completely clear of clouds.

FOG AND CIRRUS

Another cloud feature that is particularly
familiar to coastal residents is fog, a type
of stratus cloud that forms at the Earth's
surface. Although sometimes fog forms
as air rises up a mountain slope, vertical
motion is not required to produce a fog.
Fogs over the ocean, for example, tend
to form when moist air over warm ocean
water is transported over colder water. If
contact with the cooler water lowers the
temperature of the air to the dew point
temperature, fog will form without the
need to transport the air vertically. Rising in a plume is an efficient way for air
to cool. In contrast, fog-forming air
requires a longer period of time to cool
to a point that promotes the condensation of cloud drops—especially if the air
is dry. That is why fogs tend to form more
frequently near or over bodies of water,
where the air is more likely to be humid.

Thin wisps of cirrus cloud are another
common feature of fair-weather days, one
that may come and go without an accom-
panying front. As noted below, however,

LEFT: Thin wisps of cirrus adorn a deep blue sky. Such small cirrus clouds are not uncommon on fair-weather days, when they may simply reflect a small region of instability in the upper troposphere.

LOW CLOUDS
(Base usually below 6,500ft/2,000m)

TYPE	C CODE	PRECIPITATION	RANGE OF CLOUD BASE
STRATOCUMULUS (Sc)	6	Normally no ppn, but slight ppn possible over coasts and/or hills. Can be thick enough to hide the sun or moon.	Usually between 1,000ft (300m) and 4,500ft (1,350m) but may often be observed to 6,500ft (2,000m).
STRATUS (St)	7	Near coasts/hills, ppn can be considerable, but it may be falling from higher cloud such as Ns. Can be thick enough to hide sun/moon. However, when thin, sun/moon can be clearly visible.	Usually between the surface and 2,000ft (600m) but may sometimes be observed to 4,000ft (1,200m).
CUMULUS (Cu)	8	Light showers are possible.	Usually between 1,000ft/300m (although at weather stations substantially over 500ft/150m above sea level the base will often be less) and 5,000ft/1,500m, but may sometimes be observed to 6,500ft/2,000m. After initial formation, a rise in temperature often leads to a rise in cloud base.
CUMULONIMBUS (Cb)	9	Always reported when showers/thunderstorms/hail occurs. Squally winds are also common.	Usually between 2,000ft and 5,000ft (600–1,500m), but may sometimes be lower, or as high as 6,500ft (2,000m).

a thin layer of cirrostratus that thickens through the day could be a warning of an approaching storm. Furthermore, stormy weather is an important source of moisture for cirrus clouds, meaning cirrus occurring during fair weather could also be a remnant of a storm that has previously dissipated. Cirrus are composed of small ice crystals and may reside near 30,000ft (9,000m) in the mid-latitudes to as high as 50,000ft (15,000m) in the tropics.

SEVERE CONVECTIVE STORMS

Cumulus clouds are characterized by plumes of air rising in a column. The vertical rising of buoyant air in clouds is commonly referred to as convection, and the most vigorous form of it, producing the tallest clouds, occurs in the cumulonimbus clouds associated with severe convective storms. To understand how these storms become so violent, and how their cumulonimbus clouds attain great heights, it is necessary to appreciate where the energy comes from to generate the enormous buoyancy.

Convective plumes in a cumulus cloud will continue to rise if they remain warmer than their surroundings. The cloudy air cools as it rises and expands, which is fundamental to the formation of the cloud. And it is the condensation that generates the cloud that also helps to maintain its buoyancy. This is because energy is released every time a molecule of water vapor condenses and joins the liquid in a cloud drop. This release of energy heats the air in the cloud, which generates buoyancy by making the cloudy air warmer than the air outside the cloud. This means that the height of a convective cell in the core of a cumulonimbus cloud will be determined both by the temperature of the air and the amount of water vapor available at the base of the cloud to fuel condensation on the way up.

For every gram of water condensed in a cloud, approximately 600 calories of energy is released. This energy is exactly equal to the amount of energy that was required to evaporate one gram of water from the ocean, or wherever the water was released into the atmosphere. Evaporation requires energy to separate the molecules of vapor from the pull of the surrounding water molecules in the body of water from which they are evaporating. This process of evaporating water at the Earth's surface condensing in clouds is a principal source of energy for the atmosphere. While the original source of energy to drive the weather is solar energy, only a small fraction of solar energy directly heats the atmosphere. Most light from the sun passes through the atmosphere and heats the surface. This energy is later transferred to the atmosphere through evaporation and condensation.

The process of convection that describes the dynamics of plumes in cumulus clouds will yield precipitating cumulonimbus clouds if the atmosphere is sufficiently unstable and there is sufficient heat and moisture available at the base of the cloud. As buoyant plumes rise, turbulent, chaotic motions in the increasingly violent updraft cause cloud drops to collide and coalesce into larger drops. When the mass of the biggest drops grows large enough that the downward pull of gravity overcomes the vertical lift of the updraft, the cloud will begin to rain, and the cumulus cloud becomes cumulonimbus.

CUMULONIMBUS

Individual cumulonimbus clouds may expand to several thousand feet (1,000m) across and perhaps 15,000–20,000ft (6,000m) tall. Such clouds produce localized, relatively brief, showers. But towering cumulonimbus also function as the basic elements of much larger and more severe convective storms. Cumulus convection is the dominant cloud-forming mechanism in the tropics, where cumulus cells can range from shallow trade cumulus of a few hundred feet tall to the greatest cumulonimbus in the world, reaching 40,000–50,000ft (12,000–14,000m) tall. Severe convective storms also frequent the mid-latitude continents during the summer months. In North America, where warm, moist air from the Gulf of Mexico meets cold Arctic air in the summer, very violent convective storms are produced, including those that yield destructive tornados.

The basic form of a severe convective storm is the cumulonimbus with a vertically oriented convecting tower at its core. However, individual cumulonimbus clouds do not occur in isolation in a severe convective storm. These storms occur in regions where cool, dry air exists in a layer above a mass of warm humid air near ground level. Convection in cumulonimbus clouds is the

mechanism by which the atmosphere adjusts to this highly unstable condition. Heat and moisture are transported vertically in the cloud and the atmosphere is progressively stabilized. Under highly unstable conditions this may require many cumulonimbus clouds clustered together in storms that can span hundreds to thousands of square miles, or even hundreds of thousands of square miles in the case of the largest severe convective storms.

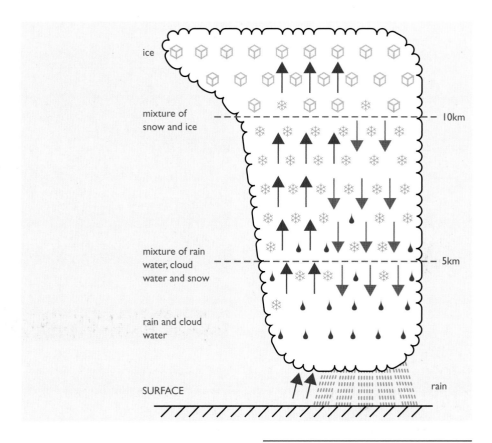

MATURE CUMULONIMBUS CLOUD

Buoyant updrafts drive the growth of cumulonimbus clouds, while heavy rain and evaporation drive downdrafts in parts of the cloud. At the base of the cloud is rain and cloud water. A mixture of cloud water, rainwater, and snow occurs where the temperature is at the freezing point of water at around 3 miles altitude (5km). Snow and ice is found higher in the cloud at temperatures of between -20 and -40 degrees Celsius. The coldest, highest portion of the cloud is composed of ice crystals.

The tops of severe convective storms frequently reach the lower boundary of the stratosphere where temperature begins to increase with height. Only upon reaching this point does the rising air within the cumulonimbus clouds lose buoyancy. The cloud drops and ice crystals being lofted in the convective cores spread out near the top of the cumulonimbus to form a thick layer of precipitating altocumulus and nimbostratus, known as an anvil cloud. The entire cloud system, including several cumulonimbus towers attached to an anvil cloud, is classified scientifically as cumulonimbus incus.

Extending away from the anvil cloud may be thinner cirrus cloud. The outflow of moisture from deep convective clouds is an important source of ice for thin cirrus clouds. Cirrus layers generated from severe storms may persist long after the storm has dissipated.

SUPERCELL STORMS

The most energetic of the mid-latitude convective storms is a supercell storm, in which many cumulonimbus cells occupy its core and together produce a broad anvil cloud. The entire mass of cloud will

move as a single propagating storm. At the heart of a supercell are violent thunderstorms with strong updrafts and heavy rains. A significant amount of this rain will be generated within the anvil cloud. The falling rain will then act to drag air down with it in a downdraft. Some of the rain will evaporate on the way down, which can accelerate the downdraft. Supercell storms frequently display some distinctive cloud features associated with downdrafts—"mamma" clouds appear as protuberances on the underside of the anvil where individual downdrafts reach the base of the cloud. Stronger downdrafts ahead of the convective

ABOVE: Storm clouds rising. An anvil begins to spread out atop a rising cumulonimbus cloud. Smaller cumulus congestus clouds surround the base of the storm and a neighboring cumulonimbus cloud grows nearby.

core of the storm can produce a powerful gust-front at the leading edge of the storm, which often produces a "shelf" cloud, also known as an arcus cloud. Shelf clouds appear as a wedge of low cloud pushing out ahead of the storm and underneath the leading edge of the anvil.

TROPICAL CYCLONES

The most destructive of severe convective storms is the tropical cyclone. Storms of this kind are composed of complex clusters of cumulonimbus clouds with large anvil cloud structures all arranged in spiral bands rotating around a center of very low pressure. These storms—which are called hurricanes when they occur in the Atlantic Ocean and typhoons when in the western Pacific Ocean (from the Chinese for "great wind")—form away from the equator over the tropical oceans, generally between eight and twenty degrees latitude (either north or south of the equator). Tropical storms are officially classified as a cyclone or hurricane when the winds exceed 74mph (119kmph). However, the strong gradient in pressure between the outer bands of the storm and the eye at the center can produce highly damaging winds up to 200mph (320kmph) in the most powerful storms.

TROPICAL CYCLONES

Low pressure at the center draws low-level air into a tropical cyclone. At the top of the storm the winds rotate in the opposite direction to the winds at the base. Strong surface winds drive evaporation from the ocean surface to fuel outer rain bands composed of sometimes violent cumulonimbus. Anvil clouds and cirrus cloud spread out from the top of the cumulonimbus (see opposite page).

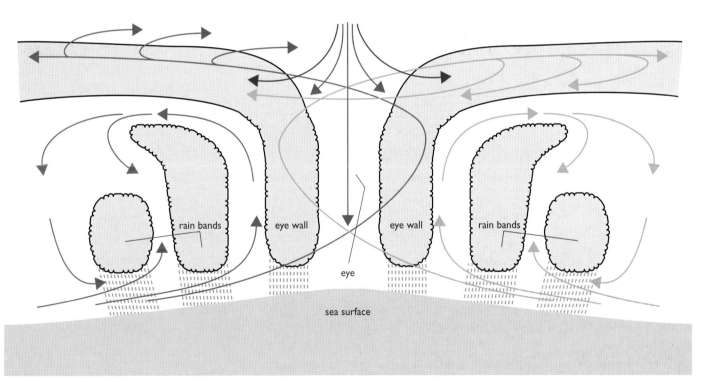

rain bands

eye wall

eye wall

rain bands

eye

sea surface

RIGHT: Hurricane Isabel coming ashore in North Carolina on September 18, 2003. Winds blew at 165mph/265kmph during Isabel's peak, but had weakened by the time of landfall. Cumulonimbus and anvil clouds extend far from the eye of the storm. Cirrus cloud is apparent as far away as southeastern Canada.

Tropical cyclones effectively act as heat engines. Underneath a cyclone, strong winds at the warm ocean surface promote rapid evaporation. The heat and moisture collected from the ocean are drawn into the updrafts of the cumulonimbus clouds, where the moisture feeds the condensation. This condensational heating helps to maintain the strong updrafts, which drive the fierce winds at the base of the storm. However, the storm must remain over very warm water in order to maintain hurricane-force winds. If the storm remains over sea water in excess of 26 degrees Celsius (80 degrees Fahrenheit), and the conditions in the upper atmosphere remain favorable for convection, a tropical cyclone can continue to churn for days. Tropical cyclones can cause major damage upon reaching land, but the winds of a cyclone generally weaken once the storm moves over the land and is cut off from the heat and moisture supplied by the ocean. The strongest cyclone observed in the Atlantic basin was Hurricane Wilma in October 2005. Sustained winds reached 185mph (298kmph) at the peak of the storm, which eventually reached Mexico, Cuba, and Florida, with winds still as high as 150mph (241kmph). The winds and related storm surge resulted in sixty-two deaths and approximately 29 billion US dollars' worth of property damage.

MID-LATITUDE CYCLONIC STORMS

The relationship between cloud-types and the structure of passing storms was first studied in detail for mid-latitude cyclonic storms, which commonly reside between thirty and sixty degrees latitude. Although they are present in the summer months, they are most frequent and energetic during the winter season. At any one time, several storms are likely to be distributed around the hemisphere, slowly making their way from west to east with the prevailing westerly winds of the so-called Roaring Forties. No two storms are exactly alike, but there are formations typical of the clouds and fronts in a mid-latitude cyclone.

Clouds and precipitation in these cyclones are organized around the fronts where warm moist air from the tropics collides with cold dry air from the polar latitudes. The warm air at a front will ride up on top of the higher density cold air. This provides the lift for producing clouds and precipitation. The general pattern of the cloud progression will be evident after one has observed the passing of several of these storms. These patterns were recognized during the systematic observations undertaken in the early twentieth century by a group of Norwegian meteorologists that has come to be known as the Bergen School. Led by Vilhelm Bjerknes, these meteorologists from the Geophysical Institute in Bergen were the first to relate the patterns of cyclonic

MEDIUM CLOUDS			
(Base usually between 6,500 and 20,000 ft although ns may be low enough to be near Earth's surface)			
TYPE	C CODE	PRECIPITATION	RANGE OF CLOUD BASE
ALTOCUMULUS (Ac)	3	Altocumulus Castellanus occasionally produces ppn. Can be thick enough to hide the sun or moon.	Usually 6,500–20,000ft (2,000–6,000m). If the height cannot reasonably be estimated, the British meterological office practice is to use a nominal height of 10,000ft/3,000m, and 15,000ft/4,500m for any Ac or As above.
ALTOSTRATUS (As)	4	Often continuous ppn reaching the ground with the sun or moon hidden. Thinner As shows sun/moon as ground glass appearance.	Altostratus may thicken with progressive lowering of the base to become Ns.
NIMBOSTRATUS (Ns)	5	Normally continuous ppn (sometimes moderate to heavy) with the sun or moon hidden. Thinner As shows sun/moon as ground glass appearance.	Usually between the surface and 10,000ft/3,000m.

storms to measurements of atmospheric pressure and the locations of fronts, thereby opening the door to modern-day weather forecasting.

As the storm advances, the warm front arrives first. The warm front is a slanted feature rising higher as it stretches out eastward, ahead of the storm. Underneath the front is a wedge of cool, dry air. Riding up the front and over the cold wedge is the warm, moist air from south of the storm. An observer hundreds of miles to the east of the storm may observe high, wispy cirrus clouds appearing in the western sky. These clouds, composed of small ice crystals, may be at an altitude of five or six miles (approximately

30,000ft/9,000m or higher). They will be the observer's earliest indication that bad weather may be approaching. The slanting ascent of warm air along the upper surface of the front means that stratus clouds are a common feature of this leading portion of the storm. The thickening of the cirrus layer into a solid layer of cirrostratus indicates that the storm is advancing. As the storm nears, the surface of the front and the base of the clouds will lower and the cloud layer will thicken further. With the cloud base below 20,000ft (6,000m), the clouds are now altostratus and may be composed of a mixture of ice, snow and water. Nimbostratus will typically arrive just ahead

A MID-LATITUDE CYCLONE

Mid-latitude cyclones ride the strong westerly winds between 30 and 60 degrees from the equator. Storm clouds form along fronts where warm moist air masses meet cold dry air masses. The warm front is preceded by cirrus clouds which eventually thicken and lower into a dense precipitating layer of nimbostratus. The cold front follows with cumulonimbus clouds followed by shallower cumulus clouds.

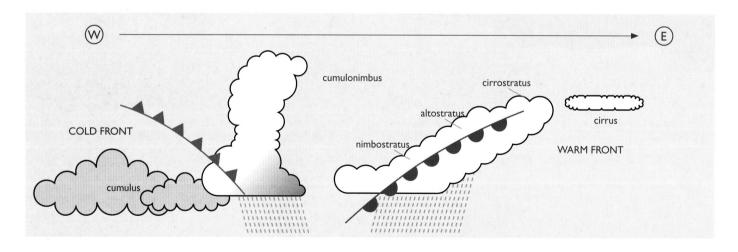

of the warm front at the surface, bringing a layer of clouds and precipitation dark enough to obscure the sun. The precipitation will be moderate to heavy and the base of nimbostratus will be 10,000ft (3,000m) or lower.

The passage of the warm front at ground level will bring a transition from the cold air mass ahead of the storm to the warm air behind the front. Conditions in the vicinity of the surface fronts may depend significantly on the heat and moisture of the warm air mass. High humidity in the air mass following the passage of the warm front will bring stratus clouds, possibly precipitating lightly—or even a fog reaching ground level. Cumulonimbus may be possible under particularly unstable conditions.

However, if the warm air mass between the warm and cold fronts is dry, the region may be predominantly clear.

Greater instability will arrive along with the cold front. Here, a cold air mass behind the front forces the warm air layer ahead of it to move upward. This upward motion may be more abrupt than the gradual slanting ascent at the warm front, touching off heavily precipitating cumulonimbus clouds close to the surface front. These clouds have a towering structure and can generate heavy rains. Air moves upward within the cumulonimbus in buoyant plumes, which can reach up to 30,000ft (9,000m)—or as high as the cirrus clouds at the leading edge of the storm. At a sharp cold front, several narrow bands

of cumulonimbus may form. Although precipitation at the surface may be in the form of rain, the upper portions of these clouds will certainly be composed of snow and ice. Conditions will generally improve rapidly after the cold front and its rain bands have passed. Clear blue skies and cool dry air may be expected after the storm.

If enough humidity is present in the air behind the cold front, which is common over the ocean, a large expanse of low cumulus clouds may move through for a while. Cumulus clouds have low bases, some 1,000–5,000ft (300–1,500m), and they extend vertically from a few hundred feet to 1,000ft (300m). Some cumulus clouds may lightly drizzle.

HIGH CLOUDS (Base usually 20,000ft/6,000m or above)			
TYPE	**C CODE**	**PRECIPITATION**	**RANGE OF CLOUD BASE**
CIRRUS (Ci)	0	None. Halo may occur. Dense patches may veil or hide the sun.	Usually 20,000–40,000ft (6,000–12,000m).
CIRROCUMULUS (Cs)	1	None. Position of sun or moon can usually be seen.	If at a non-aviation station the height cannot reasonably be estimated, it is standard practice to use a nominal height of 25,000ft,/7,600m and 35,000ft/10,700m for any higher cloud.
CIRROSTRATUS (Cs)	2	None. Halo often occurs. Outline of the sun normally visible.	Cs may thicken to become As (see MEDIUM CLOUDS).

OTHER CLOUDS

NACREOUS CLOUDS

Clouds above the troposphere are rare, and when they do occur they tend to be too thin to observe directly. At the poles, however, clouds do tend to form in the stratosphere with some regularity. These clouds are called nacreous, or "polar stratospheric," and they generally occur at 50,000–80,000ft (15,000–24,000m). Their composition is commonly combinations of either nitric acid and water or sulfuric acid and water. Polar stratospheric clouds are generally hard to observe because of their remote location and high altitude, but they became the subject of more intense scientific scrutiny when a link was discovered between polar stratospheric clouds and the chemical reaction leading to the destruction of ozone in the springtime stratosphere over Antarctica. Man-made compounds are known to cause the seasonal Antarctic ozone hole. The chemical reactions that lead to rapid ozone destruction, however, can only occur on the surfaces of ice particles comprising polar stratospheric clouds.

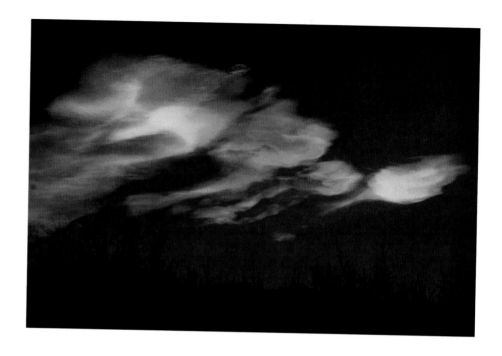

NOCTILUCENT CLOUDS

Nocilucent clouds are very thin clouds that form in the polar mesosphere at an altitude of about 50 miles (80km). These clouds are very thin and cannot be observed when overhead, or when the sun is iluminating the sky. However, when present on the horizon they can be viewed under twilight conditions, that is just before sunrise or after sunset. Because of the high altitude of the noctilucent clouds, they may still be receiving direct sunlight and an observer on the ground will see that sunlight reflecting off of the clouds. The

ABOVE: Nacreous clouds, also known as polar stratospheric clouds, are not commonly observed because of the remote locations and high altitudes where they form. These clouds are also called "mother-of-pearl clouds" in reference to their vivid iridescent appearance.

name "noctilucent" is derived from the Latin for "night-shining" because of this property of mesospheric clouds. Noctilucent clouds typically appear as silvery-blue filaments; sometimes the filaments appear orderly, but often they appear as a complex weave.

Noctilucent clouds are extremely difficult to observe, and as a result there is still

ABOVE: Noctilucent clouds (meaning "night shining") form high in the mesosphere and are seen only on the horizon at twilight. Noctilucent clouds typically appear as blue filaments against a dark sky. During daylight, the sky is too bright for these clouds to be seen.

much to be learned about how they form. Nevertheless, it is known that they form only during the summer months in each hemisphere and that northern ones are generally brighter and more frequent.

HAZE

As described previously, a cloud begins to form when an air mass has cooled to the dew point temperature and some of the molecules of water vapor have slowed enough to bond with existing drops of liquid water. However, at the dew point temperatures observed in the real atmosphere, the attractive forces between a few vapor molecules trying to bond together to form a new drop are still not strong enough to hold the drop together. If this is so, then how do new drops form in a young cloud? The answer lies in the fact that at the heart of every cloud drop is a particle. Given an object to collect on, ideally a soluble substance, such as salt, the combined attractive force of the particle and a small number of vapor molecules is sufficient to bond with additional vapor molecules.

On a humid day, a thick haze may be present and provides a glimpse of how particles in the atmosphere help contribute to cloud formation. Haze is composed of particles suspended in the atmosphere that act to reduce visibility by reflecting or "scattering" sunlight. These particles may be natural, such as dust, sea salt, volcanic emissions, or organic materials released by plants. Or they may be artificial, such as soot and other particulate pollutants emitted from vehicles and industrial processes. Even if a hazy air mass has not cooled all the way to its dew point temperature, water vapor molecules may accumulate on haze particles to form a thin film of water around the particle. These particles are termed "humidified" and will reduce the visibility even more than it would be reduced if the air mass contained only dry particles. This reduced visibility from humidified particles is most noticeable on humid days, and when there is a significant amount of pollution in the air.

OPTICAL PHENOMENA

Alongside their more typical manifestations, clouds are occasionally responsible for coloring our sky with a variety of startling optical phenomena. Among the most common of these are rainbows and halos, produced by the bending and bouncing of the sun's rays among the liquid drops and ice crystals of clouds. Although such phenomena have dazzled sky-watchers for countless generations, a proper understanding was not achieved until the nature of sunlight was discovered in the seventeenth century.

RAINBOWS

Perhaps the most common of the cloud-related optical phenomena is the rainbow. Next time you see a rainbow,

note that as you face the rainbow, the sun is behind you and a rain-cloud is before you—the rainbow is a quality of the sunlight that has been reflected off the drops and back in your direction. This observation was understood during Aristotle's day. However, the theory that passing light through a medium, in this case the water of rain

drops, could result in the bending of light—a process known as refraction—was not formulated until the Dutch physicist Willebrord Snell (1591–1626) did so in the early 1600s.

Later, a mid-seventeenth century experiment by Sir Isaac Newton, using white light and a prism, came closer to explaining the origin of rainbows. Light

---

RAINBOWS

Light from the sun, composed of colors from red to violet, passes into rain drops and is partially internally reflected before passing back out of the drop. Similar to the effect of white light travelling through a prism, the various colors of light are redirected by varying amounts in the drop. When the light emerges the colors are spread out into a rainbow.

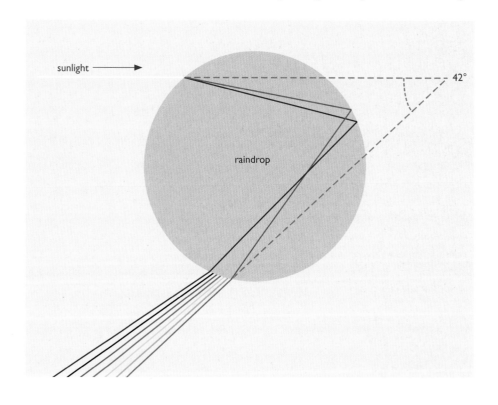

LEFT: Rainbows, such as this one appearing within an isolated shower, are seen when the rain cloud is illuminated by the sun at the observer's back.

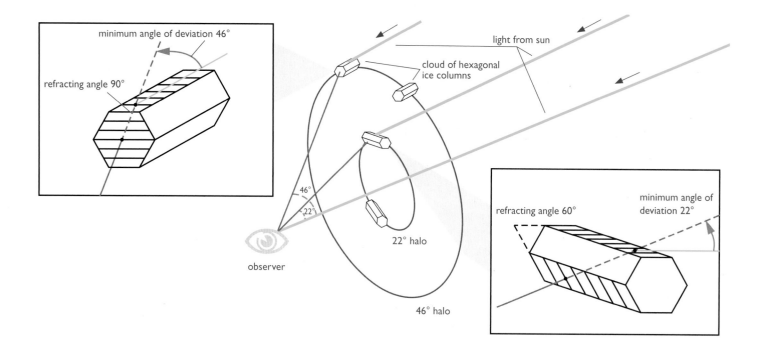

minimum angle of deviation 46°

refracting angle 90°

light from sun

cloud of hexagonal
ice columns

46°
22°

observer

22° halo

46° halo

refracting angle 60°

minimum angle of
deviation 22°

HALOS

Thin cirrus clouds composed of hexagonal columnar ice crystals can be the basis of vivid halos that ring the sun. The brighter "twenty-two-degree halo" results from the refraction of sunlight through crystals sitting perpendicular to the sun's rays, while parallel crystals produce a fainter "forty-six-degree halo."

is composed of waves known as electromagnetic waves. The light that we see with our eyes is known as "visible" light and is composed of a spectrum of electromagnetic waves corresponding to different colors, ranging from the long wavelength red to the short wavelength violet. Light of many wavelengths emanating from a single source, such as the sun, is called "white" light. Newton proved that white light is composed of a spectrum of wavelengths by passing white light through a piece of glass of varying thickness (a prism) and discovering that the light

emerged from the other side of the prism as a rainbow of different colors directed in different directions.

Raindrops have a similar effect on visible light from the sun as Newton's prism had on the light in his laboratory, although the passage of the sun's light through a raindrop is slightly more complicated. The diagram demonstrates the paths of sunrays contributing to seven colours of the rainbow. Note that the rays of different wavelength arrive at the raindrop parallel to one another, but upon entering it the angles of refraction are

LEFT: A halo bisected by several aircraft contrails. Thin layer clouds composed of columnar ice crystals often produce a halo around the sun or bright moon. This halo reveals the presence of a cloud that is otherwise barely perceptible.

different for rays of different wavelength and their paths diverge. A portion of the light travelling through the raindrop is internally reflected at the back of the raindrop and traverses it a second time. It is these rays that we observe as a rainbow when they emerge from the raindrop. The initially parallel rays of different colors exit the raindrop following divergent paths, ordered according to the familiar rainbow spectrum from violet to red.

HALOS

Halos are not nearly as common as rainbows, yet they do not require weather conditions that are particularly unusual. Halos need ice clouds, because they are a consequence of the way sunlight refracts when passing through the hexagonal-shaped, crystalline structure of ice. Next time the sky is covered by a thin layer of what appears to be very high clouds (thin cirrostratus is best) look toward the sun. If a halo is present you will see a ring around the sun of enhanced brightness that is approximately the same distance from the sun as the width of your hand held out before you with an outstretched arm. This effect is also known as the "twenty-two degree" halo because the refraction of light passing through a columnar ice crystal redirects the rays of the sun by an angle of twenty-two degrees. The twenty-two-degree halo can also be seen around a bright moon. The halo results from refraction through columnar ice crystals oriented such that the sunlight passes perpendicular to the axis of the column. This particular angle of refraction is a consequence of the hexagonal nature of the ice crystals' structure. A less frequent and less pronounced halo can sometimes be observed farther out from the sun at forty-six degrees, caused by columns oriented with their axes parallel to the sun's rays.

ADDITIONAL IMAGE CAPTIONS

CUMULUS

page 18: A lone cumulus cloud. The billowing appearance of cumulus clouds is a result of buoyant air rising within the cloud. This turbulent mixing of air leads to bulging from the sides and top of the cloud as it grows.

CUMULONIMBUS

page 30: Cumulonimbus is the tallest of the clouds. Like Cumulus clouds, cumulonimbus owes its towering form to rising, buoyant air. The tops of cumulonimbus clouds are usually high enough and cold enough for water drops to freeze, and are therefore composed of ice and snow. The frozen particles spread out at the highest reaches of the cumulonimbus to form the characteristic "anvil cloud".

page 32: Beneath the core of a cumulonimbus is a heavy downpour. Up above, and unseen, turbulent upward and downward currents buffet cloud drops and ice crystals around, causing them to coalesce into rain drops and hailstones. Overcoming the strong vertical winds, the largest of the rain drops and ice particles are drawn to the Earth by gravity, eventually emerging from the base of the cloud as rain.

page 33: Soaring cumulonimbus clouds prevent much of the sunlight that strikes them from filtering down to the cloud base. As a result, the base of a cumulonimbus can appear dark from below. Dark skies conspiring with heavy rains and blustering winds can create a disorienting environment beneath cumulonimbus clouds.

page 36–37: Flanked by cumulus congestus, a mature cumulonimbus generates a broad anvil cloud. Despite the savage conditions experienced in the heart of such clouds, cumulonimbus was the inspiration for the expression of contentment, "being on cloud nine". Clearly to be on the top of this cloud is far preferable to being inside it.

STRATUS

page 44: Stratus opacus is so-called because it is thick enough to extensively mask the light of the sun.

page 50: A valley fog slowly conceals the landcape as it settles on the valley floor. This phenomenon is caused by heavier cold air settling in a valley, with warmer air passing over the mountains above it— a form of radiation fog confined by local topography that can last for several days.

page 51: Shreds of broken stratus cloud settle on the hill slopes near taller cumulus cloud. These shreds are known as stratus fractus. This mountain-top may have previously been covered by an orographic stratus cloud that has become unstable and is beginning to grow into cumulus clouds.

STRATOCUMULUS

page 54: A deck of stratocumulus clouds is formed of numerous shallow convective clouds that have merged into a single layer. The top surface pitches and heaves with the turbulent mixing occurring in the cloud.

page 64: The distinctive texture of stratocumulus clouds is usually more apparent from above than below. Here, the visual contrasts that result from filtered sunlight reveal the complex dynamics that exist within a seemingly static layer of cloud. The appearance of some structural form is a clear indication that the cloud is stratocumulus, rather than stratus.

page 65: The cool and dark conditions beneath an overcast stratocumulus layer set a gloomy mood. This is compounded by a soaking drizzle, which sometimes forms in patches.

ALTOCUMULUS

page 66: "Mackerel sky, mackerel sky, never long wet, never long dry." A sky such as this field of altocumulus cloudlets was referred to as a "mackerel sky" by sailors for its resemblance to the scales of a mackerel fish. Such a sky commanded sailors' attention not just for its fish-like appearance; the "mackerel sky" often precedes the passage of showers.

ALTOSTRATUS

page 80: Uniform and gray, altostratus are rarely very appealing. Higher than stratus clouds, they similarly reveal the slow ascent over a wide area. Although they often appear to simply hang there unchanging, altostratus often moves in ahead of a storm, appearing to develop from a thickening and lowering layer of cirrostratus.

pages 82–83: Altostratus is at its most alluring when the sun is low in the sky. As the sun rises or sets, a featureless, leaden deck is gradually transformed into a shifting blend of hues. Moments before sunrise, this layer of altostratus is displaying a myriad shades of violet.

NIMBOSTRATUS

page 92: Nimbostratus often appear to reach all the way to the Earth's surface. If one accepts that rain is nothing more than cloud drops that have grown large enough to fall to the ground, then indeed this cloud does reach the ground. This rain is produced well above the surface where a large mass of air is slowly rising.

CIRRUS

page 102: The name "cirrus" is derived from the Latin for a lock of hair, and from the appearance of these cirrus clouds it

is clear why they are so named. When the individual fibres of cirrus are long and distinct, such as these, the cloud is classified as cirrus fibratus.

CIRROCUMULUS

page 114: The characteristically small cloudlets that make up a thin layer of cirrocumulus stratiformis are arranged into a wave pattern by instability in the blustery winds at high altitude.

CIRROSTRATUS

page 126: Cirrostratus nebulosus lacks any clearly defined features or texture, making its boundaries nearly impossible to distinguish precisely. This example is unusually dark because the sun is low in the sky. Cirrostratus is typically transparent and allows enough sun to penetrate that objects on the ground cast shadows.

OTHER CLOUDS AND PHENOMENA

page 134: The sun is ringed with a brilliant halo, courtesy of a transparent deck of cirrostratus cloud. The ice crystals in the cloud act as tiny prisms bending the light from the sun as it passes through the cloud, producing the halo. Also like a prism, the colors of the light are slightly spread giving the inner edge of the halo a red hue and the outer edge a violet hue.

Hamblyn, R. *The Invention of Clouds: How an Amateur Meteorologist Forged the Language of the Skies.* Ferrar Straus Giroux: New York, 2001.

Pretor-Pinney, G. *The Cloudspotter's Guide.* Hodder & Stoughton: London, 2006.

Rubin, L. D. and J. Duncan. *The Weather Wizard's Cloud Book.* Algonquin Books of Chapel Hill: New York, 1989.

Burroughs, W. J., B. Crowder, T. Robertson, E. Vallier-Talbot, R. Whitaker. *Weather.* (Nature Company Guides) Time-Life Books: New York, 1996.

Bohren, C. F. *Clouds in a Glass of Beer: Simple Experiments in Atmospheric Physics.* John Wiley & Sons, Inc: Hoboken, New Jersey, 1987.

Wallace, J. M. and P. V. Hobbs. *Atmospheric Science: An Introductory Survey.* Academic Press Inc: San Diego, 1977.

WEB RESOURCES:

Cloud types for observers from the U.K. Met Office, http://www.metoffice.gov.uk/publications/clouds/index.html

Glossary of Meteorology from The American Meteorological Society, http://amsglossary.allenpress.com/glossary

INDEX

ACKNOWLEDGMENTS & PICTURE CREDITS

ACKNOWLEDGMENTS

The author would like to thank Dr. Santiago Gassó for his valuable comments. The diagram on p.170 is after p.226 of John M. Wallace and Peter V. Hobbs, *Atmospheric Science: An Introductory Survey*. Academic Press: San Diego, 1977.

PICTURE CREDITS

The publisher would like to thank the following people and photographic libraries for permission to reproduce their material. Every care has been taken to trace copyright holders. However, if we have omitted anyone we apologize and will, if informed, make corrections to any future edition.

Page 2–3 Harald Edens/www.weatherscapes. com; **6–7** David Edwards/National Geographic/Getty Images; **8–9** Pekka Parviainen/Science Photo Library; **18** Natural Selection John Bracchi/Design Pics/Corbis; **20–21** Anup Shah/Getty Images; **22** Nacivet/ Getty Images; **23** Gerolf Kalt/zefa/Corbis; **24–25** Masaaki Toyoura/Getty Images; **26–27** Issac Rose/Alamy; **28–29** Stephen Wilkes/Getty Images; **30** Dr Dan Sudia/ Science Photo Library; **32** P. Manner/zefa/ Corbis; **33** Stephanie Cabrera/Getty Images; **34–35** Adam Jones/Getty Images; **36–37** Kaj R. Svensson/Science Photo Library; **38** Jim Reed/Corbis; **39** Jim Reed/Corbis; **40–41** Eddie Soloway/Getty Images; **42–43** Gene Rhoden/weatherpix.com; **44** Veikko Vasama/ Arctic Photoagency Leuku; **46–47** Laura Ciapponi/Getty Images; **48–49** Stephen Frink Collection/Alamy; **50** Royal Geographical Society/Alamy; **50–51** image100/Corbis; **52–53** Sakke Nenye/Arctic Photoagency Leuku; **54** Michael Duva/Getty Images; **56–57** Suk–Heui Park/Getty Images; **58–59** Ann Johansson/Corbis; **60–61** Martin Hospach/ Getty Images; **62–63** Image Source Black/ Getty Images; **64–65** Martti Kapanen/Arctic Photoagency Leuku; **65** Louie Psihoyos/ Getty Images; **66** Craig Tuttle/Corbis; **68–69** Kevin Schafer/Getty Images; **70** Harald Edens/ www.weatherscapes.com; **71** Harald Edens/ www.weatherscapes.com; **72–73** Veikko Vasama/Arctic Photoagency Leuku; **74–75** Pekka Parviainen/Science Photo Library; **76–77** Image Source Black/Getty Images; **78–79** Jens Lucking/Getty Images; **80** Harald Edens/www.weatherscapes.com; **82–83** Momatiuk–Eastcott/Corbis; **84–85** Jeff Foott/ Getty Images; **86–87** Clark Dunbar/Corbis; **88** Harald Edens/www.weatherscapes.com; **89** Harald Edens/www.weatherscapes.com; **90–91** Gene Rhoden/weatherpix.com; **92** Terraqua Images/Corbis; **94–95** Fernando Bengoechea/Beateworks/Corbis; **96–97** Harald Edens/www.weatherscapes.com; **98–99** image100/Corbis; **100–101** Pete Leonard/zefa/Corbis; **102** DLILLC/Corbis; **104–105** Tom Bean/Corbis; **106–107** Gene Rhoden/weatherpix.com; **108** Craig Tuttle/ Corbis; **109** Antti Saraja/Arctic Photoagency Leuku; **110–111** Matthieu Ricard/Getty Images; **112–113** Gene Rhoden/weatherpix. com; **114** Harald Edens/www.weatherscapes. com; **116–117** Harald Edens/www. weatherscapes.com; **118** Harald Edens/ www.weatherscapes.com; **119** Harald Edens/www.weatherscapes.com; **120–121** Panoramic Images/Getty Images; **122–123** Pekka Parviainen/Science Photo Library; **124–125** Harald Edens/www.weatherscapes. com; **126** image100/Corbis; **128–129** Clark Dunbar/Corbis; **130–131** Harald Edens/ www.weatherscapes.com; **132–133** Matthias Clamer/Getty Images; **134** Visuals Unlimited/ Corbis; **136–137** Michel Gounot/Godong/ Corbis; **138–139** Jorma Luhta/Arctic Photoagency Leuku; **140–141** David Hay Jones/Science Photo Library; **142–143** Harald Edens/www.weatherscapes.com; **144–145** Zen Shui/Milena Boniek/Getty Images; **148** NOAA; **150** NASA; **155** Timothy Allen/ Axiom Photographic Agency; **156** Rich Reid/ Getty Images; **157** www.image-ark.com/ Philippe Gueissaz/Arctic Photoagency Leuku; **160** L. Clarke/Corbis; **162** NASA; **166** David Hay Jones/Science Photo Library; **167** Pekka Parviainen/Science Photo Library; **169** Chip Porter/Getty Images; **171** Harald Edens/ www.weatherscapes.com.